Christabel Donatienne Ruby (Ed.)

Nokia E50

Christabel Donatienne Ruby (Ed.)

Nokia E50

Multi-band, Smartphone, Nokia

Fidel

Imprint

Publisher:
Fidel is a trademark of
International Book Market Service Ltd., 17 Rue Meldrum, Beau Bassin, 1713-01 Mauritius
Email: info@bookmarketservice.com
Website: www.bookmarketservice.com

Published in 2012

Printed in: U.S.A., U.K., Germany. This book was not produced in Mauritius.

ISBN: 978-620-0-71513-5

Contents

Articles

Nokia_E50	1
Nokia	5
Nokia_Eseries	33
Multi-band	35
Intellisync	36
Smartphone	37
Visto	61
Secure_Digital	64
ActiveSync	86

References

Article Sources and Contributors	90
Image Sources, Licenses and Contributors	92

Nokia_E50

Manufacturer	Nokia
Compatible networks	EGSM 850/900/1800/1900
Successor	Nokia E51
Form factor	Candybar
Dimensions	113 × 43.5 × 15.5 millimeters
Weight	104 g (**unknown operator: u'strong'** lb)
Operating system	S60 3rd edition on Symbian OS v9.1
CPU	ARM 9 @ 235 MHz
Memory	70 MB (internal)
Removable storage	MicroSD
Battery	BL-5C (970 mAh) or BL-6C (1150 mAh)
Display	2.06 in, 31×42 mm, 240×320 pixels, Active Matrix, 262,144 colours
Rear camera	1,280×960 pixels (1.3 megapixel), 4× digital zoom
Connectivity	USB Mass Storage via Pop-Port, Bluetooth 2.0, Infrared

The **Nokia E50** Business Device is a bar-style monoblock quad-band smartphone from Nokia announced May 18, 2006[1] as part of the Eseries, intended primarily for the corporate business market. It includes sophisticated e-mail support for Nokia's Intellisync Wireless Email, BlackBerry Connect, Visto Mobile, Activesync Mail for Exchange, Altexia as well as IMAP4. It also has the ability to view Microsoft Word, PowerPoint, and Excel attachments, and PDF documents but it cannot be used for editing these without additional apps. An application manager downloads, removes and installs both Nokia and third-party applications. Device to device synchronization is possible with Data transfer application. Features include EDGE, Bluetooth 2.0, a 1,280 × 960 pixels (1.3 megapixel) camera, a MicroSD memory-card slot, and digital music and video player functionality through RealPlayer and Flash Player. This unit does *not* support UMTS, Wi-Fi, or FM radio.

Nokia E50 in hand

It uses the third edition of the Series 60 user-interface (S60v3) and the Symbian operating system version 9.1. It is *not* binary compatible with software compiled for earlier versions of the Symbian operating system.

Although this phone is marked down as the smartphone, it is really, in fact a feature phone.

Versions

Check firmware

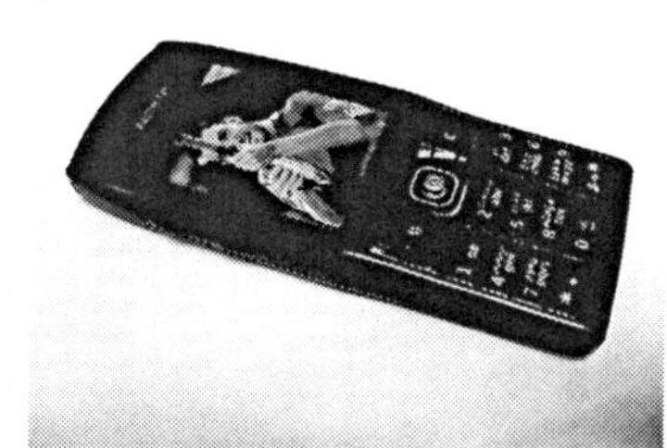

Nokia E50 Metal Black

*#0000#

V 07.36.0.0[2]

- Nokia model E50-1 / RM-170 (with camera)
- Nokia model E50-2 / RM-171 (without camera)
- Nokia model E50 Metal Black, available in selected markets.[3]

Specifications sheet

Feature	Specification
Form factor	Candybar / Monoblock
Operating System	Symbian OS (9.1) + Series 60 v3
GSM frequencies	850/900/1800/1900 MHz
GPRS	Yes, class 10
EDGE (EGPRS)	Yes, class 10
3G	yes
UMTS/WCDMA	No
WLAN/WiFi	No
PG.R.S. support	Yes
Main screen	Active Matrix, 262 144 colours, 240 × 320 pixels, adjustable brightness
Camera (optional)	1,280 × 960 pixels (1.3 megapixel), 4x digital zoom
Video recording	Yes 176 × 144 or 128 × 96, 15 frames per second, up to one hour
Voice recording	Yes
Multimedia Messaging	Yes
Video calls	No
Push to talk	Yes (Push to Talk over Cellular – PoC)
Java support	Yes, MIDP 2.0
Built-in memory	70 MB
Memory card slot	Yes, MicroSD/TransFlash max. 2 GB
Hot Swappable Memory card slot	Yes
Bluetooth	Yes v2.0 + EDR
Infrared	Yes
USB Mass Storage	Yes, via Pop-Port
Data cable support	Yes
Browser	WAP 2.0 XHTML / HTML. Comes standard with a full Nokia Mini Map Browser
Email	Yes
Music player	Yes, stereo
Radio FM	No
Video Player	Yes
Stereo Speakers	No
Ringtones	Yes, Polyphonic, Monophonic, MP3, True Tones
Vibrate	Yes
HandsFree (HF) speakerphone	Yes
Offline/Flight mode	Yes
Battery	BL-5C (970 mAh) or BL-6C (1,150 mAh)
Talk time	6.8 / 8 hours
Standby time	9 days / 10 days
Weight	104 grams

Dimensions	113 × 43.5 × 15.5 millimeters
Availability	Q3 2006
Else	Nokia PC Suite

See also

- Smartphone
- Nokia Eseries

References

[1] "The smallest of Nokia Eseries, the Nokia E50 business device for mobile professionals" (http://press.nokia.com/PR/200605/1051798_5.html) (Press release). Nokia. 2006-05-18. . Retrieved 2006-05-28.
[2] Nokia Firmware Versions (http://www.developer.nokia.com/Community/Wiki/Nokia_Firmware_Versions)
[3] Nokia E50 Key Features (http://europe.nokia.com/A4158064)

External links

- Nokia E50 – Product Page (http://europe.nokia.com/phones/e50)
- Nokia E50 – Resource Information (http://www.forum.nokia.com/info/sw.nokia.com/id/ddd8018c-e10c-4327-bc7f-ca37a41af912/E50.html)
- Nokia E50 – Device Details (http://www.forum.nokia.com/devices/E50)
- Bluetooth Qualification Program specifications (http://qualweb.bluetooth.org/Template2.cfm?LinkQualified=QualifiedProducts&Details=Yes&ProductID=3614)
- Repair Video (http://www.youtube.com/watch?v=oUkEfg0UAsQ)

Reviews, photos and videos

- Nokia E50 – Review by Mobile-Review (http://www.mobile-review.com/review/nokia-e50-en.shtml)
- Nokia E50 – Review by phoneArena.com (http://www.phonearena.com/htmls/readreviews.php?id=1473&page=0)
- Nokia E50 – Review by GSM Arena (http://www.gsmarena.com/nokia_e50-review-109.php), Video (http://www.gsmarena.com/nokia_e50-video-1566.php)
- Nokia E50 – Review by About-nokia.com (http://www.about-nokia.com/blog/index.php?itemid=362)
- Nokia E50 – Preview by All About Symbian (http://www.allaboutsymbian.com/features/item/Nokia_E50_Hands_on_Preview.php)
- Nokia E50 – Review by CNET: UK (http://reviews.cnet.co.uk/mobiles/0,39030106,49287030,00.htm)
- Nokia E50 – Review by Tech2 India (http://www.tech2.com/india/reviews/smart-mobile-phones/nokia-e50/3764/0)
- Nokia E50 – Review by Steve Punter's Southern Ontario Cell Phone Page (http://www.arcx.com/sites/NokiaE50.htm)
- Nokia E50 – Unofficial Live Photos of 'Metal Black' model (http://nokiae50.info/articles/live-photos-of-nokia-e50-metal-black)
- Nokia E50 – Official Photos of 'Metal Black' model (http://www.nokia.com.my/BaseProject/Sites/MYEN_49531/CDA/Categories/Home/downloads/photos/_Content/_Static_Files/e50blkcombo_high.zip)
- Nokia E50 – Official Photos (http://www.nokia.com/A4136017?category=e50)
- Nokia E50 – Official Video (http://www.europe.nokia.com/A4158135)

Nokia

NOKIA	
Type	Public company
Traded as	OMX: NOK1V [1], NYSE: NOK [2], FWB: NOA3 [3]
Industry	Telecommunications Internet Computer software
Founded	Tampere, Finland, Russian Empire (1865) incorporated in Nokia (1871)
Founder(s)	Fredrik Idestam Leo Mechelin
Headquarters	Espoo, Finland,European Union
Area served	Worldwide
Key people	Jorma Ollila (Chairman) Stephen Elop (President & CEO) Timo Ihamuotila (CFO) Kai Öistämö (CDO) Henry Tirri (CTO)
Products	Mobile phones Smartphones Mobile computers Networks (See products listing)
Services	Maps and navigation, music, messaging and media Software solutions (See services listing)
Revenue	€38.65 billion (2011)[4]
Operating income	€-1.073 billion (2011)[4]
Net income	€-1.164 billion (2011)[4]
Total assets	€36.20 billion (2011)[4]
Total equity	€11.87 billion (2011)[4]
Employees	135,949 (2011)[4]
Divisions	Mobile Solutions Mobile Phones Markets
Subsidiaries	Nokia Siemens Networks Navteq Symbian Vertu Qt Development Frameworks
Website	Nokia.com [5]

Nokia Corporation (Finnish pronunciation: [ˈnɔkiɑ]) (OMX: NOK1V [1], NYSE: NOK [2], FWB: NOA3 [3]) is a Finnish multinational communications corporation that is headquartered in Keilaniemi, Espoo, a city neighbouring Finland's capital Helsinki.[6] Nokia manufactures mobile electronic devices, mostly mobile telephones and other devices related to communications, and in converging Internet and communications industries, with over 132,000 employees in 120 countries, sales in more than 150 countries and global annual revenue of over €42 billion and operating profit of €2 billion as of 2010.[4] It was the world's largest manufacturer of mobile phones in 2011, with global device market share of 23% in the second quarter.[7] Nokia's estimated share of the converged mobile device market was 31% in the fourth quarter, compared with 38% in the third quarter of 2010.[4] Nokia produces mobile devices for every major market segment and protocol, including GSM, CDMA, and W-CDMA (UMTS). Nokia offers Internet services such as applications, games, music, maps, media and messaging through its Ovi platform. Nokia's joint venture with Siemens, Nokia Siemens Networks produces telecommunications network equipment, solutions and services.[8] Nokia also provides free-of-charge digital map information and navigation services through its wholly owned subsidiary Navteq.[9]

Nokia has sites for research and development, manufacture and sales in several countries; as of December 2010, Nokia had R&D presence in 16 countries and employed 35,870 people in research and development, representing approximately 27% of the group's total workforce.[4]

Nokia is a public limited-liability company listed on the Helsinki, Frankfurt, and New York stock exchanges, [10] and plays a very large role in the economy of Finland, accounting for about a third of the market capitalization of the Helsinki Stock Exchange (OMX Helsinki) in 2007.[11]

The Nokia brand, valued at $25 billion, is listed as the 14th most valuable global brand in the Interbrand/*BusinessWeek* Best Global Brands list of 2011.[12] It is the 14th ranked brand corporation in Europe (as of 2011),[13] the 8th most admirable *Network and Other Communications Equipment* company worldwide in Fortune's World's Most Admired Companies list of 2011,[14] and the world's 143th largest company as measured by revenue in Fortune Global 500 list of 2011.[15] In July 2010, Nokia reported a drop in profits by 40%,[16] which turned into an operating loss of €487 million in Q2 2011.[17] In the global smartphone rivalry,[18] Nokia held the 3rd place in 2Q2011, trailing behind Samsung and Apple.[19] [20]

On 11 February 2011 Nokia announced a partnership with Microsoft; all Nokia smartphones introduced since then were to run under Microsoft's Windows Phone (WP) operating system. On 26 October 2011 Nokia unveiled its first Windows Phone handsets, the WP7.5 Lumia 710 and 800.[21]

History

The Nokia House, Nokia's head office located by the Gulf of Finland in Keilaniemi, Espoo, was constructed between 1995 and 1997. It is the workplace of more than 1,000 Nokia employees.[22]

Pre-telecommunications era

Fredrik Idestam, founder of Nokia. Statesman Leo Mechelin, co-founder of Nokia.

The predecessors of the modern Nokia were the Nokia Company (Nokia Aktiebolag), Finnish Rubber Works Ltd (Suomen Gummitehdas Oy) and Finnish Cable Works Ltd (Suomen Kaapelitehdas Oy).[23]

Nokia's history started in 1865 when mining engineer Fredrik Idestam established a groundwood pulp mill on the banks of the Tammerkoski rapids in the town of Tampere, in southwestern Finland in Russian Empire and started manufacturing paper.[24] In 1868, Idestam built a second mill near the town of Nokia, fifteen kilometres (nine miles) west of Tampere by the Nokianvirta river, which had better resources for hydropower production.[25] In 1871, Idestam, with the help of his close friend statesman Leo Mechelin, renamed and transformed his firm into a share company, thereby founding the Nokia Company, the name it is still known by today.[25]

Toward the end of the 19th century, Mechelin's wishes to expand into the electricity business were at first thwarted by Idestam's opposition. However, Idestam's retirement from the management of the company in 1896 allowed Mechelin to become the company's chairman (from 1898 until 1914) and sell most shareholders on his plans, thus realizing his vision.[25] In 1902, Nokia added electricity generation to its business activities.[24]

Industrial conglomerate

In 1898, Eduard Polón founded Finnish Rubber Works, manufacturer of galoshes and other rubber products, which later became Nokia's rubber business.[23] At the beginning of the 20th century, Finnish Rubber Works established its factories near the town of Nokia and they began using Nokia as its product brand.[26] In 1912, Arvid Wickström founded Finnish Cable Works, producer of telephone, telegraph and electrical cables and the foundation of Nokia's cable and electronics businesses.[23] At the end of the 1910s, shortly after World War I, the Nokia Company was nearing bankruptcy.[27] To ensure the continuation of electricity supply from Nokia's generators, Finnish Rubber Works acquired the business of the insolvent company.[27] In 1922, Finnish Rubber Works acquired Finnish Cable Works.[28] In 1937, Verner Weckman, a sport wrestler and Finland's first Olympic Gold medalist, became President of Finnish Cable Works, after 16 years as its Technical Director.[29] After World War II, Finnish Cable Works supplied cables to the Soviet Union as part of Finland's war reparations. This gave the company a good foothold for later trade.[29]

The three companies, which had been jointly owned since 1922, were merged to form a new industrial conglomerate, Nokia Corporation in 1967 and paved the way for Nokia's future as a global corporation.[30] The new company was involved in many industries, producing at one time or another paper products, car and bicycle tires, footwear (including rubber boots), communications cables, televisions and other consumer electronics, personal computers, electricity generation machinery, robotics, capacitors, military communications and equipment (such as the SANLA M/90 device and the M61 gas mask for the Finnish Army), plastics, aluminium and chemicals.[22] Each business unit had its own director who reported to the first Nokia Corporation President, Björn Westerlund. As the president of the Finnish Cable Works, he had been responsible for setting up the company's first electronics department in 1960, sowing the seeds of Nokia's future in telecommunications.[31]

Eventually, the company decided to leave consumer electronics behind in the 1990s and focused solely on the fastest growing segments in telecommunications.[32] Nokian Tyres, manufacturer of tires, split from Nokia Corporation to form its own company in 1988[33] and two years later Nokian Footwear, manufacturer of rubber boots, was founded.[26] During the rest of the 1990s, Nokia divested itself of all of its non-telecommunications businesses.[32]

Telecommunications era

The seeds of the current incarnation of Nokia were planted with the founding of the electronics section of the cable division in 1960 and the production of its first electronic device in 1962: a pulse analyzer designed for use in nuclear power plants.[31] In the 1967 fusion, that section was separated into its own division, and began manufacturing telecommunications equipment. A key CEO and subsequent Chairman of the Board was *vuorineuvos* Björn "Nalle" Westerlund (1912–2009), who founded the electronics department and let it run at a loss for 15 years.

Networking equipment

In the 1970s, Nokia became more involved in the telecommunications industry by developing the Nokia DX 200, a digital switch for telephone exchanges. The DX 200 became the workhorse of the network equipment division. Its modular and flexible architecture enabled it to be developed into various switching products.[34] In 1984, development of a version of the exchange for the Nordic Mobile Telephony network was started.[35]

For a while in the 1970s, Nokia's network equipment production was separated into *Telefenno*, a company jointly owned by the parent corporation and by a company owned by the Finnish state. In 1987, the state sold its shares to Nokia and in 1992 the name was changed to Nokia Telecommunications.

In the 1970s and 1980s, Nokia developed the Sanomalaitejärjestelmä ("Message device system"), a digital, portable and encrypted text-based communications device for the Finnish Defence Forces.[36] The current main unit used by the Defence Forces is the Sanomalaite M/90 (SANLA M/90).[37]

First mobile phones

The technologies that preceded modern cellular mobile telephony systems were the various "0G" pre-cellular mobile radio telephony standards. Nokia had been producing commercial and some military mobile radio communications technology since the 1960s, although this part of the company was sold some time before the later company rationalization. Since 1964, Nokia had developed VHF radio simultaneously with Salora Oy. In 1966, Nokia and Salora started developing the ARP standard (which stands for Autoradiopuhelin, or *car radio phone* in English), a car-based mobile radio telephony system and the first commercially operated public mobile phone network in Finland. It went online in 1971 and offered 100% coverage in 1978.[40]

The Mobira Cityman 150, Nokia's NMT-900 mobile phone from 1989 (left), compared to the Nokia 1100 from 2003.[38] The Mobira Cityman line was launched in 1987.[39]

In 1979, the merger of Nokia and Salora resulted in the establishment of Mobira Oy. Mobira began developing mobile phones for the NMT (Nordic Mobile Telephony) network standard, the first-generation, first fully automatic cellular phone system that went online in 1981.[41] In 1982, Mobira introduced its first car phone, the Mobira Senator for NMT-450 networks.[41]

Nokia bought Salora Oy in 1984 and now owning 100% of the company, changed the company's telecommunications branch name to Nokia-Mobira Oy. The Mobira Talkman, launched in 1984, was one of the world's first transportable phones. In 1987, Nokia introduced one of the world's first handheld phones, the Mobira Cityman 900 for NMT-900 networks (which, compared to NMT-450, offered a better signal, yet a shorter roam). While the Mobira Senator of 1982 had weighed 9.8 kg (**unknown operator: u'strong'** lb) and the Talkman just under 5 kg (**unknown operator: u'strong'** lb), the Mobira Cityman weighed only 800 g (**unknown operator: u'strong'** oz) with the battery and had a price tag of 24,000 Finnish marks (approximately €4,560).[39] Despite the high price, the first phones were almost snatched from the sales assistants' hands. Initially, the mobile phone was a "yuppie" product and a status symbol.[22]

Nokia's mobile phones got a big publicity boost in 1987, when Soviet leader Mikhail Gorbachev was pictured using a Mobira Cityman to make a call from Helsinki to his communications minister in Moscow. This led to the phone's nickname of the "Gorba".[39]

In 1988, Jorma Nieminen, resigning from the post of CEO of the mobile phone unit, along with two other employees from the unit, started a notable mobile phone company of their own, Benefon Oy (since renamed to GeoSentric).[42] One year later, Nokia-Mobira Oy became Nokia Mobile Phones.

Involvement in GSM

Nokia was one of the key developers of GSM (Global System for Mobile Communications),[43] the second-generation mobile technology which could carry data as well as voice traffic. NMT (Nordic Mobile Telephony), the world's first mobile telephony standard that enabled international roaming, provided valuable experience for Nokia for its close participation in developing GSM, which was adopted in 1987 as the new European standard for digital mobile technology.[44] [45]

Nokia delivered its first GSM network to the Finnish operator Radiolinja in 1989.[46] The world's first commercial GSM call was made on 1 July 1991 in Helsinki, Finland over a Nokia-supplied network, by then Prime Minister of Finland Harri Holkeri, using a prototype Nokia GSM phone.[46] In 1992, the first GSM phone, the Nokia 1011, was launched.[46] [47] The model number refers to its launch date, 10 November.[47] The Nokia 1011 did not yet employ Nokia's characteristic ringtone, the Nokia tune. It was introduced as a ringtone in 1994 with the Nokia 2100 series.[48]

GSM's high-quality voice calls, easy international roaming and support for new services like text messaging (SMS) laid the foundations for a worldwide boom in mobile phone use.[46] GSM came to dominate the world of mobile telephony in the 1990s, in mid-2008 accounting for about three billion mobile telephone subscribers in the world, with more than 700 mobile operators across 218 countries and territories. New connections are added at the rate of 15 per second, or 1.3 million per day.[49]

Personal computers and IT equipment

In the 1980s, Nokia's computer division Nokia Data produced a series of personal computers called MikroMikko.[50] MikroMikko was Nokia Data's attempt to enter the business computer market. The first model in the line, MikroMikko 1, was released on 29 September 1981,[51] around the same time as the first IBM PC. However, the personal computer division was sold to the British ICL (International Computers Limited) in 1991, which later became part of Fujitsu.[52] MikroMikko remained a trademark of ICL and later Fujitsu. Internationally the MikroMikko line was marketed by Fujitsu as the ErgoPro.

The Nokia Booklet 3G mini laptop.

Fujitsu later transferred its personal computer operations to Fujitsu Siemens Computers, which shut down its only factory in Espoo, Finland (in the Kilo district, where computers had been produced since the 1960s) at the end of March 2000,[53] [54] thus ending large-scale PC manufacturing in the country. Nokia was also known for producing very high quality CRT and early TFT LCD displays for PC and larger systems application. The Nokia Display Products' branded business was sold to ViewSonic in 2000.[55] In addition to personal computers and displays, Nokia used to manufacture DSL modems and digital set-top boxes.

Nokia re-entered the PC market in August 2009 with the introduction of the Nokia Booklet 3G mini laptop.[56]

Challenges of growth

In the 1980s, during the era of its CEO Kari Kairamo, Nokia expanded into new fields, mostly by acquisitions. In the late 1980s and early 1990s, the corporation ran into serious financial problems, a major reason being its heavy losses by the television manufacturing division and businesses that were just too diverse.[57] These problems, and a suspected total burnout, probably contributed to Kairamo taking his own life in 1988. After Kairamo's death, Simo Vuorilehto became Nokia's Chairman and CEO. In 1990–1993, Finland underwent severe economic depression,[58] which also struck Nokia. Under Vuorilehto's management, Nokia was severely overhauled. The company responded by streamlining its telecommunications divisions, and by divesting itself of the television and PC divisions.[59]

Probably the most important strategic change in Nokia's history was made in 1992, however, when the new CEO Jorma Ollila made a crucial strategic decision to concentrate solely on telecommunications.[32] Thus, during the rest of the 1990s, the rubber, cable and consumer electronics divisions were gradually sold as Nokia continued to divest itself of all of its non-telecommunications businesses.[32]

As late as 1991, more than a quarter of Nokia's turnover still came from sales in Finland. However, after the strategic change of 1992, Nokia saw a huge increase in sales to North America, South America and Asia.[60] The exploding worldwide popularity of mobile telephones, beyond even Nokia's most optimistic predictions, caused a logistics crisis in the mid-1990s.[61] This prompted Nokia to overhaul its entire logistics operation.[62] By 1998, Nokia's focus on telecommunications and its early investment in GSM technologies had made the company the world's largest mobile phone manufacturer.[60] Between 1996 and 2001, Nokia's turnover increased almost fivefold from 6.5 billion euros to 31 billion euros.[60] Logistics continues to be one of Nokia's major advantages over its rivals, along with greater economies of scale.[63] [64]

Recent history

Product releases

Nokia released its first touch screen phone, the Nokia 7710, which was a huge success. In May 2007, Nokia announced that its Nokia 1100 handset, launched in 2003,[38] with over 200 million units shipped, was the best-selling mobile phone of all time and the world's top-selling consumer electronics product.[65] In November 2007, Nokia announced and released the Nokia N82, its first Nseries phone with Xenon flash. At the Nokia World conference in December 2007, Nokia announced their "Comes With Music" program: Nokia device buyers are to receive a year of complimentary access to music downloads.[66] The service became commercially available in the second half of 2008.

Reduction in size of Nokia mobile phones

Nokia Productions was the first ever mobile filmmaking project directed by Spike Lee. Work began in April 2008, and the film premiered in October 2008.[67]

In 2008, Nokia released the Nokia E71 which was marketed to directly compete with the other BlackBerry-type devices offering a full "qwerty" keyboard and cheaper prices. Nokia announced in August 2009 that they will be selling a high-end Windows-based mini laptop called the Nokia Booklet 3G.[56] On 2 September 2009, Nokia launched two new music and social networking phones, the X6 and X3.[68] The Nokia X6 features 32GB of on-board memory with a 3.2" finger touch interface and comes with a music playback time of 35 hours. The Nokia X3 is a first series 40 Ovi Store-enabled device. The X3 is a music device that comes with stereo speakers, built-in FM radio, and a 3.2 megapixel camera. On 10 September 2009, Nokia unveiled a new handset, the 7705 Twist, a phone with a sports square shape that swivels open to reveal a full QWERTY keypad.[69] The new mobile, which will be available exclusively through Verizon Wireless, features a 3 megapixel camera, web browsing, voice commands and weighs around 3.44 ounces (**unknown operator: u'strong'** g).

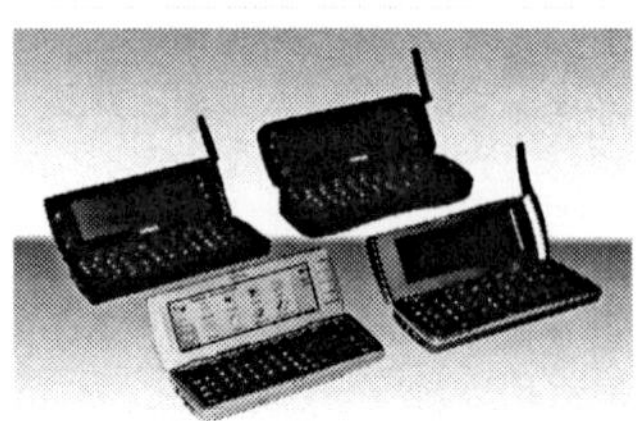

Evolution of the Nokia Communicator. Models 9000, 9110, 9210, 9300 and 9500 shown.

Plant movements

Nokia opened its Komárom, Hungary mobile phone factory on 5 May 2000.[70]

In March 2007, Nokia signed a memorandum with Cluj County Council, Romania to open a new plant near the city in Jucu commune.[71] [72] [73] Moving the production from the Bochum, Germany factory to a low wage country created an uproar in Germany.[74] [75] Nokia recently moved its North American Headquarters to Sunnyvale.

Reorganizations

In April 2003, the troubles of the networks equipment division caused the corporation to resort to similar streamlining practices on that side, including layoffs and organizational restructuring.[76] This diminished Nokia's public image in Finland,[77] [78] and produced a number of court cases and an episode of a documentary television show critical of Nokia.[79]

On February 2006, Nokia and Sanyo announced a memorandum of understanding to create a joint venture addressing the CDMA handset business. But in June, they announced ending negotiations without agreement. Nokia also stated its decision to pull out of CDMA research and development, to continue CDMA business in selected markets.[80] [81] [82]

In June 2006, Jorma Ollila left his position as CEO to become the chairman of Royal Dutch Shell[83] and to give way for Olli-Pekka Kallasvuo.[84] [85]

In May 2008, Nokia announced on their annual stockholder meeting that they want to shift to the Internet business as a whole. Nokia no longer wants to be seen as the telephone company. Google, Apple and Microsoft are not seen as natural competition for their new image but they are considered as major important players to deal with.[86]

In November 2008, Nokia announced it was ceasing mobile phone distribution in Japan.[87] Following early December, distribution of Nokia E71 is cancelled, both from NTT docomo and SoftBank Mobile. Nokia Japan retains global research & development programs, sourcing business, and an MVNO venture of Vertu luxury phones, using docomo's telecommunications network.

Acquisitions

For a more comprehensive list, see List of acquisitions by Nokia.

On 22 September 2003, Nokia acquired Sega.com, a branch of Sega which became the major basis to develop the Nokia N-Gage device.[88]

The Nokia E55 from the business segment of the Eseries range

On 16 November 2005, Nokia and Intellisync Corporation, a provider of data and PIM synchronization software, signed a definitive agreement for Nokia to acquire Intellisync.[89] Nokia completed the acquisition on 10 February 2006.[90]

On 19 June 2006, Nokia and Siemens AG announced the companies would merge their mobile and fixed-line phone network equipment businesses to create one of the world's largest network firms, Nokia Siemens Networks.[91] Each company has a 50% stake in the infrastructure company, and it is headquartered in Espoo, Finland. The companies predicted annual sales of €16 bn and cost savings of €1.5 bn a year by 2010. About 20,000 Nokia employees were transferred to this new company.

On 8 August 2006, Nokia and Loudeye Corp. announced that they had signed an agreement for Nokia to acquire online music distributor Loudeye Corporation for approximately US $60 million.[92] The company has been developing this into an online music service in the hope of using it to generate handset sales. The service, launched on 29 August 2007, is aimed to rival iTunes. Nokia completed the acquisition on 16 October 2006.[93]

In July 2007, Nokia acquired all assets of Twango, the comprehensive media sharing solution for organizing and sharing photos, videos and other personal media.[94] [95]

In September 2007, Nokia announced its intention to acquire Enpocket, a supplier of mobile advertising technology and services.[96]

In October 2007, pending shareholder and regulatory approval, Nokia bought Navteq, a U.S.-based supplier of digital mapping data, for a price of $8.1 billion.[9] [97] Nokia finalized the acquisition on 10 July 2008.[98]

In September, 2008, Nokia acquired OZ Communications, a privately held company with approximately 220 employees headquartered in Montreal, Canada.[99]

On 24 July 2009, Nokia announced that it will acquire certain assets of cellity, a privately owned mobile software company which employs 14 people in Hamburg, Germany.[100] The acquisition of cellity was completed on 5 August 2009.[101]

On 11 September 2009, Nokia announced the acquisition of "certain assets of Plum Ventures, Inc, a privately held company which employed approximately 10 people with main offices in Boston, Massachusetts. Plum will complement Nokia's Social Location services".[102]

On 28 March 2010, Nokia announced the acquisition of Novarra, the mobile web browser firm from Chicago. Terms of the deal were not disclosed.Novarra is a privately held company based in Chicago, IL and provider of a mobile browser and service platform and has more than 100 employees.[103]

On 10 April 2010, Nokia announced its acquisition of MetaCarta, whose technology was planned to be used in the area of local search, particularly involving location and other services. Financial details of acquisition were not disclosed.[104]

Curtailments

Amid falling sales, Nokia posted a loss of 368 million euros for Q2 2011, while in Q2 2010 had still a profit of 227 million euros. On September 2011, Nokia has announced it will lose another 3,500 jobs worldwide, including the closure of its Cluj factory in Romania.[105]

On February 8, 2012 Nokia Corp. said to cut around 4,000 jobs at smartphone manufacturing plants in Europe by the end of 2012 to move assembly closer to component supplier in Asia. It plans to cut 2,300 of the 4,400 jobs in Hungaria, 700 out of 1,000 jobs in Mexico, and 1,000 out of 1,700 factory jobs in Finland.[106]

Operating systems

Originally Nokia phones had a custom Nokia OS operating system developed specifically for Nokia mobile phones.

The first Nseries device, the N90, utilised the older Symbian OS 8.1 mobile operating system, as did the N70. Subsequently Nokia switched to using SymbianOS 9 for all later Nseries devices (except the N72, which was based on the N70). Newer Nseries devices incorporate newer revisions of SymbianOS 9 that include Feature Packs. The N800, N810 and N900 are as of July 2010 the only Nseries devices to not use Symbian OS. They use the Linux-based Maemo.[107]

Nokia stated that Maemo would be developed alongside Symbian.

Maemo has since (Maemo "6" and beyond) merged with Intel's Moblin, and become MeeGo, which will continue to be developed for mobile devices.

The Nokia N8 is the first device to function on the Symbian^3 mobile operating system.

Nokia revealed that the N8 will be the last device in its flagship N-series devices to ship with Symbian OS.[108] [109]

Instead, Nokia will use Microsoft Windows Phone for its high-end flagship devices, and revealed the Nokia N9 will function on the MeeGo mobile operating system.

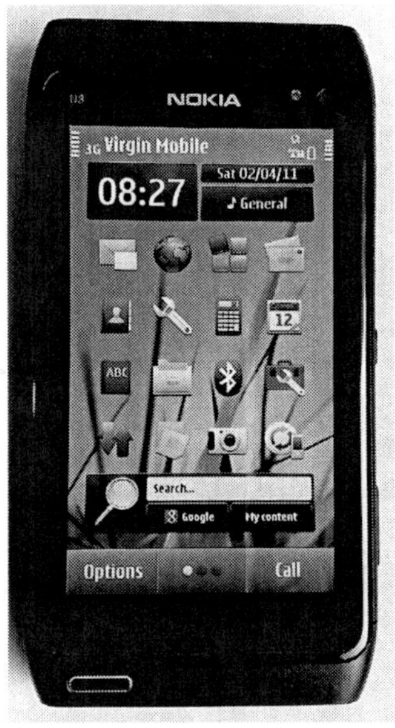

Photograph taken using Nokia N82 The Nokia N8 smartphone is the world's first Symbian^3 device, and the first camera phone to ever feature a 12 megapixel autofocus lens.

Alliance with Microsoft

On 11 February 2011, Nokia's CEO Stephen Elop, a former Microsoft employee, unveiled a new strategic alliance with Microsoft, and announced it would replace Symbian and MeeGo with Microsoft's Windows Phone operating system[110] [111] except for mid-to-low-end devices, which would continue to run under Symbian. Nokia was also to invest into the Series 40 platform and release a single MeeGo product in 2011.[112]

This news was not well received by consumers, and has contributed to the decline in the stock price by 11%.[113]

As part of the restructuring plan, Nokia planned to reduce spending on research and development, instead customising and enhancing the software line for Windows Phone 7.[114] Nokia's "applications and content store" (Ovi) becomes integrated into the Windows Phone Marketplace, and Nokia Maps is at the heart of Microsoft's Bing and AdCenter. Microsoft provides developer tools to Nokia to replace the Qt framework, which is not supported by Windows Phone 7 devices.[115]

Symbian becomes described as a "franchise platform" with Nokia planning to sell 150 million Symbian devices after the alliance was set up. MeeGo emphasis is on longer-term exploration, with plans to ship "a MeeGo-related product" later in 2012. Microsoft's search engine, Bing becomes the search engine for all Nokia phones. Nokia also gets some level of customisation on WP7.[116]

After this announcement, Nokia's share price fell about 14%, its biggest drop since July 2009.[117]

As Nokia was the largest mobile phone manufacturer worldwide at the time,[7] it is suggested the alliance would make Microsoft's Windows Phone 7 a stronger contender against Android and iOS.[115] In June 2011 Nokia was overtaken by Apple as the world's biggest smartphone maker by volume.[118] In August 2011 Chris Weber, head of Nokia's subsidiary in the U.S., stated "*The reality is if we are not successful with Windows Phone, it doesn't matter what we do (elsewhere).*" He further added "*North America is a priority for Nokia (...) because it is a key market for Microsoft.*"[119]

Corporate affairs

Corporate structure

Divisions

Since 1 July 2010, Nokia comprises three business groups: **Mobile Solutions**, **Mobile Phones** and **Markets**.[120] The three units receive operational support from the **Corporate Development Office**, led by Kai Öistämö, which is also responsible for exploring corporate strategic and future growth opportunities.[120]

On 1 April 2007, Nokia's Networks business group was combined with Siemens's carrier-related operations for fixed and mobile networks to form Nokia Siemens Networks, jointly owned by Nokia and Siemens and consolidated by Nokia.[121]

Mobile Solutions

Mobile Solutions is responsible for Nokia's portfolio of smartphones and mobile computers, including the more expensive multimedia and enterprise-class devices. The team is also responsible for a suite of internet services under the Ovi brand, with a strong focus on maps and navigation, music, messaging and media.[120] This unit is led by Anssi Vanjoki, along with Tero Ojanperä (for Services) and Alberto Torres (for MeeGo Computers).[120]

The Nokia N900, a Maemo 5 Linux based mobile Internet device and touchscreen smartphone from Nokia's Nseries portfolio.

Alberto Torres has stepped down.

Mobile Phones

The Nokia E90, a Symbian smartphone from Nokia's Eseries portfolio.

Mobile Phones is responsible for Nokia's portfolio of affordable mobile phones, as well as a range of services that people can access with them, headed by Mary T. McDowell.[120] This unit provides the general public with mobile voice and data products across a range of devices, including high-volume, consumer oriented mobile phones. The devices are based on GSM/EDGE, 3G/W-CDMA and CDMA cellular technologies.

In the first quarter of 2006 Nokia sold over 15 million MP3 capable mobile phones, which means that Nokia is not only the world's leading supplier of mobile phones and digital cameras (as most of Nokia's mobile telephones feature digital cameras, it is also believed that Nokia has recently overtaken Kodak in camera production making it the largest in the world), Nokia is now also the leading supplier of digital audio players (MP3 players), outpacing sales of devices such as the iPod from Apple. At the end of the year 2007, Nokia managed to sell almost 440 million mobile phones which accounted for 40% of all global mobile phones sales.[122] By 2010, Nokia's market share in the mobile phone market had dropped to 32.6% (453 million phones).[123]

Anssi Vanjoki resigned a few days before Nokia World 2010 and under new leadership team Jo Harlow will look into the affairs of Smartphones portfolio.

On 27 April 2011, The Register reported that Nokia is secretly developing a new operating system called Meltemi aiming at the low-end market. It is believed it will be replacing the S30 and S40 operating systems. Due to low-end market customers' demand of having smartphone features in their feature phone, the OS will include some features exclusive to high-end smartphones.

Markets

Markets is responsible for Nokia's supply chains, sales channels, brand and marketing functions of the company, and is responsible for delivering mobile solutions and mobile phones to the market. The unit is headed by Niklas Savander.[120]

Subsidiaries

Nokia has several subsidiaries, of which the two most significant as of 2009 are Nokia Siemens Networks and Navteq.[120] Other notable subsidiaries include, but are not limited to Vertu, a British-based manufacturer and retailer of luxury mobile phones; Qt Software, a Norwegian-based software company, and OZ Communications, a consumer e-mail and instant messaging provider.

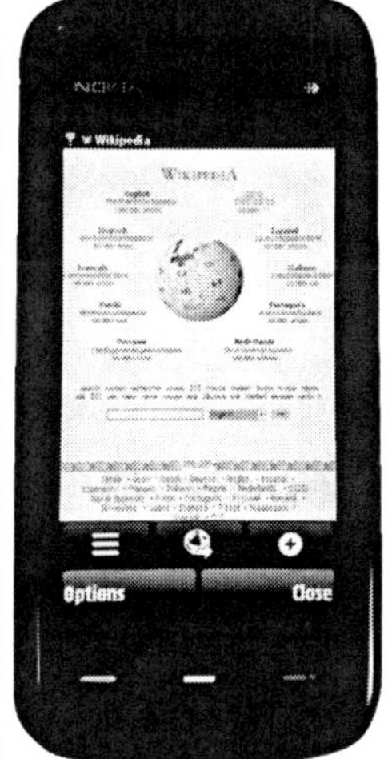

The Nokia 5800 XpressMusic, a touchscreen smartphone and portable entertainment device which emphasizes music and multimedia playback.

Until 2008 Nokia was the major shareholder in Symbian Limited, a software development and licensing company that produced Symbian OS, a smartphone operating system used by Nokia and other manufacturers. In 2008 Nokia acquired Symbian Ltd and, along with a number of other companies, created the Symbian Foundation to distribute the Symbian platform royalty free and as open source.

Nokia Siemens Networks

Nokia Siemens Networks (previously Nokia Networks) provides wireless and fixed network infrastructure, communications and networks service platforms, as well as professional services to operators and service providers.[120] Nokia Siemens Networks focuses in GSM, EDGE, 3G/W-CDMA and WiMAX radio access networks; core networks with increasing IP and multiaccess capabilities; and services.

On 19 June 2006 Nokia and Siemens AG announced the companies are to merge their mobile and fixed-line phone network equipment businesses to create one of the world's largest network firms, called Nokia Siemens Networks.[91] The Nokia Siemens Networks brand identity was subsequently launched at the 3GSM World Congress in Barcelona in February 2007.[124] [125]

As of March 2009, Nokia Siemens Networks serves more than 600 operator customers in more than 150 countries, with over 1.5 billion people connected through its networks.[126]

On 22 August 2011 Nokia Siemens became embroiled in a scandal related to the use and abuse of surveillance systems delivered to the Bahrain government by one of its former business units, Nokia Siemens Intelligence Solutions (NSIS). The spy gear in Bahrain was sold by Siemens AG (SIE), and maintained by Nokia Siemens Networks and NSN's divested unit, Trovicor GmbH. The sale and maintenance contracts were also confirmed by Ben Roome, a Nokia Siemens spokesman based in Farnborough, England. The system was reportedly used as the investigative tool of choice to gather information about political dissidents—and silence them. Companies such as Nokia and Nokia Siemens are free to sell such equipment almost anywhere. For the most part, the U.S. and European countries lack export controls to deter the use of such systems for repression, as was the case in Bahrain were at least 30 people were killed during the 2011 uprising. Many Western nations actively support the export of these systems of repression, e.g. to countries that are home to some of the U.S. Navy's Fleet. Monitoring centers, as the systems are called, are sold around the globe by Nokia Siemens and its competitors, such as Israel-based Nice Systems Ltd. (NICE), and Verint Systems Inc. (VRNT), headquartered in Melville, New York. They form the heart of so-called lawful interception surveillance systems. By the end of 2007, the Nokia Siemens Intelligence Solutions unit had more than 90 systems installed in 60 countries.[127] Besides Bahrain, several other Middle Eastern nations that cracked down on uprisings this year—including Egypt, Syria and Yemen—also purchased monitoring centers from the chain of businesses now known as Trovicor. Trovicor equipment plays a surveillance role in at least 12 Middle Eastern and North African nations. Trovicor's precursor, which started in 1993 as the voice- and data-recording unit of Siemens, in 2007 became part of Nokia Siemens Networks, the world's second biggest maker of wireless communications equipment. NSN, a 50-50 joint venture with Espoo, Finland-based Nokia Oyj (NOK1V), sold the unit, known as Intelligence Solutions, in March 2009. The new owners, Guernsey-based Perusa Partners Fund 1 LP, renamed the business Trovicor, coined from the Latin and Esperanto words for find and heart, according to the company's website. According to NSN the elevated risk of human rights abuses was a major reason for NSN's exiting the monitoring-center business. In Bahrain, officials routinely used the NSIS surveillance systems as a basis for the arrest and torture of political opponents; legally the monitoring technology is to be only used by order of legal authorities such as judges and prosecutors. According to local regulations, every Bahraini phone and Internet operator must provide the state with the ability to monitor communications. Phone companies also must track the location of phones within a 164-foot (50-meter) radius, the rules say. NSN and Trovicor's status as exclusive provider in Bahrain continued at least through 2009. That period of more than two years coincides with the dates of text messages used to interrogate scores of political detainees. Authorities used messages that dated as far back as the mid-2000s, even in recent interrogations.[128]

Navteq

Navteq is a Chicago, Illinois-based provider of digital map data and location-based content and services for automotive navigation systems, mobile navigation devices, Internet-based mapping applications, and government and business solutions.[120] Navteq was acquired by Nokia on 1 October 2007.[9] Navteq's map data is part of the Nokia Maps online service where users can download maps, use voice-guided navigation and other context-aware web services.[120] Nokia Maps is part of the Ovi brand of Nokia's Internet based online services.

Corporate governance

The control and management of Nokia is divided among the shareholders at a general meeting and the Group Executive Board (left),[129] under the direction of the Board of Directors (right).[130] The Chairman and the rest of the Group Executive Board members are appointed by the Board of Directors. Only the Chairman of the Group Executive Board can belong to both, the Board of Directors and the Group Executive Board. The Board of Directors' committees consist of the Audit Committee,[131] the Personnel Committee[132] and the Corporate Governance and Nomination Committee.[133] [134]

The operations of the company are managed within the framework set by the Finnish Companies Act,[135] Nokia's Articles of Association[136] and Corporate Governance Guidelines,[137] and related Board of Directors adopted charters.

Group Executive Board (January 2011) [129]
Stephen Elop (Chairman), b. 1963 President, CEO and Group Executive Board Chairman of Nokia Corporation since 21 September 2010 Joined Nokia on 21 September 2010
Esko Aho, b. 1954 Executive Vice President, Corporate Relations and Responsibility Joined Nokia 2008, Group Executive Board member since 2009 Former Prime Minister of Finland (1991–1995)
Jerri DeVard, b. 1958 Executive Vice President, Chief Marketing Officer Joined Nokia 2011, Group Executive Board member since 2011

Timo Ihamuotila, b. 1966 Executive Vice President, Chief Financial Officer With Nokia 1993–1996, rejoined 1999, Group Executive Board member since 2007
Mary T. McDowell, b. 1964 Executive Vice President, Mobile Phones Joined Nokia 2004, Group Executive Board member since 2004
Dr. Tero Ojanperä, b. 1966 Executive Vice President, Services, Mobile Solutions Joined Nokia 1990, Group Executive Board member since 2005
Niklas Savander, b. 1962 Executive Vice President, Markets Joined Nokia 1997, Group Executive Board member since 2006
Alberto Torres, b. 1965 Executive Vice President, MeeGo Computers, Mobile Solutions Joined Nokia 2004, Group Executive Board member since 2009
Juha Äkräs, b. 1965 Executive Vice President, Human Resources Joined Nokia 1993, Group Executive Board member since 2010
Dr. Kai Öistämö, b. 1964 Executive Vice President, Chief Development Officer Joined Nokia 1991, Group Executive Board member since 2005

Board of Directors [130]
Jorma Ollila (Chairman), b. 1950 Board member since 1995, Chairman of the Board of Directors since 1999 Chairman of the Board of Directors of Royal Dutch Shell PLC
Dame Marjorie Scardino (Vice Chairman), b. 1947 Board member since 2001 Chairman of the Corporate Governance and Nomination Committee, Member of the Personnel Committee Chief Executive Officer and member of the Board of Directors of Pearson PLC
Lalita D. Gupte, b. 1948 Board member since 2007, Member of the Audit Committee Non-executive Chairman of the ICICI Venture Funds Management Co Ltd.
Dr. Bengt Holmström, b. 1949 Board member since 1999 Paul A. Samuelson Professor of Economics at Massachusetts Institute of Technology, joint appointment at the MIT Sloan School of Management
Dr. Henning Kagermann, b. 1947 Board member since 2007, Member of the Personnel Committee Former CEO and Chairman of the Executive Board of SAP AG
Per Karlsson, b. 1955 Board member since 2002, Independent Corporate Advisor Chairman of the Personnel Committee, Member of the Corporate Governance and Nomination Committee
Isabel Marey-Semper, b. 1967 Board member since 2009, Member of the Audit Committee Chief Financial Officer, EVP in charge of strategy of PSA Peugeot Citroën
Risto Siilasmaa, b. 1966 Board member since 2008, Member of the Audit Committee Founder and Chairman of F-Secure

Keijo Suila, b. 1945 Board member since 2006, Member of the Audit Committee

Former corporate officers

Chief Executive Officers		Chairmen of the Board of Directors [138]			
Björn Westerlund	1967–1977	Lauri J. Kivekäs	1967–1977	Simo Vuorilehto	1988–1990
Kari Kairamo	1977–1988	Björn Westerlund	1977–1979	Mika Tiivola	1990–1992
Simo Vuorilehto	1988–1992	Mika Tiivola	1979–1986	Casimir Ehrnrooth	1992–1999
Jorma Ollila	1992–2006	Kari Kairamo	1986–1988		
Olli-Pekka Kallasvuo	2006–2010				

International presence

In 2010 Nokia had 132,000 employees in 120 countries, sales in more than 150 countries, global annual revenue of over €42 billion, and operating profit of €2 billion.[4] It was the world's largest manufacturer of mobile phones in 2011, with global device market share of 23% in the second quarter.[7] Nokia's estimated share of the converged mobile device market was 31% in the fourth quarter, compared with 38% in the third quarter of 2010.[4]

Nokia has sites for research and development, manufacture and sales in several countries; as of December 2010, Nokia had R&D presence in 16 countries and employed 35,870 people in research and development, representing approximately 27% of the group's total workforce.[4] The Nokia Research Center, founded in 1986, is Nokia's industrial research unit consisting of about 500 researchers, engineers and scientists;[139] [140] it has sites in seven countries: Finland, China, India, Kenya, Switzerland, the United Kingdom and the United States.[141] Besides its research centers, in 2001 Nokia founded (and owns) INdT – Nokia Institute of Technology, a R&D institute located in Brazil.[142] Nokia operates a total of 9 manufacturing facilities[10] located at Salo, Finland; Manaus, Brazil; Cluj, Romania; Beijing and Dongguan, China; Komárom, Hungary; Chennai, India; Reynosa, Mexico; and Masan, South Korea.[143] [71] Nokia's factory in Cluj was seized by the Romanian government in November 2011 to prevent a sale of the assets, after Nokia had accumulated a tax liability of US$ 10 million.[144] Nokia's industrial design department is headquartered in Soho in London, UK with significant satellite offices in Helsinki, Finland and Calabasas, California in the US.

Nokia is a public limited-liability company listed on the Helsinki, Frankfurt, and New York stock exchanges.[10] Nokia plays a very large role in the economy of Finland; it is by far the largest Finnish company, accounting for about a third of the market capitalization of the Helsinki Stock Exchange (OMX Helsinki) in 2007, a unique situation for an industrialized country.[145] It is an important employer in Finland and several small companies have grown into large ones as its partners and subcontractors.[146] In 2009 Nokia contributed 1.6% to Finland's GDP, and accounted for about 16% of Finland's exports in 2006.[147]

In February 2012 Nokia announced that it was cutting 4,000 factory jobs in Finland, Hungary and Mexico (more than half of the 7,100 jobs at the three factories affected) and moving smartphone assembly to existing facilities in South Korea and China[148] .

—===Logos===

Past

Nokia Company logo. Founded in Tampere in 1865, incorporated in Nokia, FinlandNokia in 1871.

The brand logo of Nokian FootwearFinnish Rubber Works, founded in Helsinki in 1898.Logo from 1965 to 1966.

The Nokia Corporation "arrows" logo, used before the "Connecting People" logo.

NOKIA
CONNECTING PEOPLE

Nokia introduced its "Connecting People" advertising slogan, coined by Ove Strandberg "HS Archives" (in Finnish). Helsingin Sanomat. 1 June 2003. . Retrieved 14 May 2008. and used since 1992. "NOKIA I Connecting Pople 1992 Vector Logo (AI EPS)". HDicon.com. . Retrieved 17 October 2010.This earlier version of the slogan used Times RomanTimes Roman SC (Small Caps) font.Pitkänen, Juhani (Nokia's Art Director) (3 September 2007). "Nokia Strategic Marketing, Brand Identity". Nokia Corporation. . Retrieved 14 May 2008.

Present

NOKIA
Connecting People

Nokia's current logo used since 2006, "NOKIA I Connecting Pople new Vector Logo (AI EPS)". HDicon.com. . Retrieved 17 October 2010. with the redesigned "Connecting People" slogan.This slogan uses Nokia's proprietary 'Nokia Sans' font, designed by Erik Spiekermann. "Erik Spiekermann – Furniture, Designs & Home Decor". Design Within Reach. . Retrieved 7 January 2010.

Nokia Siemens Networks logo. Founded in 2007.

Stock

Nokia, a public limited liability company, is the oldest company listed under the same name on the Helsinki Stock Exchange (since 1915).[22] Nokia's shares are also listed on the Frankfurt Stock Exchange (since 1988) and New York Stock Exchange (since 1994).[10] [22]

In 1 June 2011 Nokia shares dropped to their lowest in more than 13 years. Nokia shares fell as much as 10 percent, extending their previous day's by 18 percent fall.[154]

For fiscal Q2 2011 ending in June 2011, Nokia reported a net loss of 492 million EUR, despite a 430 million EUR payment from Apple. Nokia cited decline in its mobile phone business as the primary cause of the loss.[155]

Corporate culture

The Nokia House, Nokia's head office in Keilaniemi, Espoo, Finland.

Nokia's official corporate culture manifesto, *The Nokia Way*, emphasises the speed and flexibility of decision-making in a flat, networked organization, although the corporation's size necessarily imposes a certain amount of bureaucracy.[156]

The official business language of Nokia is English. All documentation is written in English, and is used in official intra-company spoken communication and e-mail.

Until May 2007, the *Nokia Values* were Customer Satisfaction, Respect, Achievement, and Renewal. In May 2007, Nokia redefined its values after initiating a series of discussions worldwide as to what the new values of the company should be. Based on the employee suggestions, the new values were defined as: Engaging You, Achieving Together, Passion for Innovation and Very Human.[156]

Online services

.mobi and the Mobile Web

Nokia was the first proponent of a Top Level Domain (TLD) specifically for the Mobile Web and, as a result, was instrumental in the launch of the .mobi domain name extension in September 2006 as an official backer.[157] [158] Since then, Nokia has launched the largest mobile portal, Nokia.mobi [159], which receives over 100 million visits a month.[160] It followed that with the launch of a mobile Ad Service [161] to cater to the growing demand for mobile advertisement.[162]

Ovi

Nokia Ovi logo.

Ovi, announced on 29 August 2007, is the name for Nokia's "umbrella concept" Internet services.[163] Centered on Ovi.com, it is marketed as a "personal dashboard" where users can share photos with friends, download music, maps and games directly to their phones and access third-party services like Yahoo's Flickr photo site. It has some significance in that Nokia is moving deeper into the world of Internet services, where head-on competition with Microsoft, Google and Apple is inevitable.[164]

The services offered through Ovi include the Ovi Store (Nokia's application store), the Nokia Music Store, Nokia Maps, Ovi Mail, the N-Gage mobile gaming platform available for several S60 smartphones, Ovi Share, Ovi Files, and Contacts and Calendar.[165] The Ovi Store, the Ovi application store was launched in May 2009.[166] Prior to opening the Ovi Store, Nokia integrated its software Download! store, the stripped-down MOSH repository and the widget service WidSets into it.[167]

On 23 March 2010, Nokia announced launch of its online magazine called the *Nokia Ovi*. The 44-page magazine contains articles on products by Nokia, what Ovi stands for, tips and tricks on the usage of Nokia mini laptop Booklet 3G, latest reviews of mobile applications, news about the mobile maker's services and apps such as Ovi maps, files and mail. Users can download the magazine as a PDF or view it online from the Nokia website.[168]

My Nokia

Nokia offers a free personalised service to Nokia owners called My Nokia (located at my.nokia.com).[169] Registered My Nokia users can get free services as follows:

- Tips & tricks alerts through web, e-mail and also mobile text message.
- My Nokia Backup: A free online backup service for mobile contacts, calendar logs and also various other files. This service needs GPRS connection.
- Ringtones, wallpapers, screensavers, games and other things can be downloaded free of cost.

Comes With Music

In 2007 Nokia set up their "Nokia Comes With Music" service, in partnership with Universal Music Group International, Sony BMG, Warner Music Group, EMI, and hundreds of Independent labels and music aggregators, to allow 12, 18, or 24 months of unlimited free-of-charge music downloads with the purchase of a Nokia Comes With Music edition phone. Files could be downloaded on mobile devices or personal computers, and kept permanently.[66]

In January 2011 Nokia withdrew this program in 27 countries, due to its failure to gain traction with customers or mobile network operators; existing subscribers could continue to download until their contracts ended. The service continued to be offered in China, India, Indonesia, Brazil, Turkey and South Africa where take-up had been better.[170]

Nokia Messaging

On 13 August 2008 Nokia launched a beta release of "Nokia Email service", a push e-mail service, since incorporated into Nokia Messaging.[171]

Nokia Messaging operates as a centralised, hosted service that acts as a proxy between the Nokia Messaging client and the user's e-mail server. The phone does not connect directly to the e-mail server, but instead sends e-mail credentials to Nokia's servers.[172] IMAP is used as the protocol to transfer emails between the client and the server.

Controversy

NSN's provision of intercept capability to Iran

In 2008, Nokia Siemens Networks, a joint venture between Nokia and Siemens AG, reportedly provided Iran's monopoly telecom company with technology that allowed it to intercept the Internet communications of its citizens to an unprecedented degree.[173] The technology reportedly allowed it to use deep packet inspection to read and even change the content of everything from "e-mails and Internet phone calls to images and messages on social-networking sites such as Facebook and Twitter". The technology "enables authorities to not only block communication but to monitor it to gather information about individuals, as well as alter it for disinformation purposes," expert insiders told *The Wall Street Journal*. During the post-election protests in Iran in June 2009, Iran's Internet access was reported to have slowed to less than a tenth of its normal speeds, and experts suspected this was due to the use of the interception technology.[174]

The joint venture company, Nokia Siemens Networks, asserted in a press release that it provided Iran only with a 'lawful intercept capability' "solely for monitoring of local voice calls". "Nokia Siemens Networks has not provided any deep packet inspection, web censorship or Internet filtering capability to Iran," it said.[175]

In July 2009, Nokia began to experience a boycott of their products and services in Iran. The boycott was led by consumers sympathetic to the post-election protest movement and targeted at those companies deemed to be collaborating with the Islamic regime. Demand for handsets fell and users began shunning SMS messaging.[176]

Lex Nokia

In 2009, Nokia heavily supported the passing of a law in Finland that allows companies to monitor their employees' electronic communications in cases of suspected information leaking.[177] Contrary to rumors, Nokia denied that the company would have considered moving its head office out of Finland if laws on electronic surveillance were not changed.[178] The law was enacted, but with strict requirements for implementation of its provisions. As of 2010, the law has become a dead letter; no corporation has implemented it. The Finnish media dubbed the name *Lex Nokia* for this law, named after the Finnish copyright law (the so-called *Lex Karpela*) a few years back.

Nokia–Apple patent dispute

In October 2009, Nokia filed a lawsuit against Apple Inc. in the U.S. District Court of Delaware citing Apple infringed on 10 of its patents related to wireless communication including data transfer.[179] Apple was quick to respond with a countersuit filed in December 2009 accusing Nokia of 11 patent infringements. Apple's General Counsel, Bruce Sewell went a step further by stating, "Other companies must compete with us by inventing their own technologies, not just by stealing ours." This resulted in an ugly spat between the two telecom majors with Nokia filing another suit, this time with the U.S. International Trade Commission (ITC), alleging Apple of infringing its patents in "virtually all of its mobile phones, portable music players, and computers."[180] Nokia went on to ask the court to bar all U.S. imports of the Apple products including the iPhone, Mac and the iPod. Apple countersued by filing a complaint with the ITC in January 2010, the details of which are yet to be confirmed.[179]

In June 2011, Apple settled with Nokia and agreed to an estimated one time payment of $600 million and royalties to Nokia.[181] The two companies also agreed on a cross-licensing patents for some of their patented technologies.[182][183]

Environmental record

Electronic products such as cell phones impact the environment both during production and after their useful life when they are discarded and turned into electronic waste. Nokia is listed in Greenpeace's Guide to Greener Electronics that scores leading electronics manufacturers according to their policies on sustainability, climate and energy and how green their products are. In November 2011 Nokia ranked 3rd out of 15 listed electronics companies, falling two places due to its weaker performance on the Energy criteria and scoring 4.9/10.[184]

All of Nokia's mobile phones are free of toxic polyvinyl chloride (PVC) since the end of 2005 and all new models of mobile phones and accessories launched in 2010 are on track to be free of brominated compounds, chlorinated flame retardants and antimony trioxide.[184]

Nokia's voluntary take-back programme to recycle old mobile phones spans 84 countries with almost 5,000 collection points.[185] However, the recycling rate of Nokia phones was only 3–5% in 2008, according to a global consumer survey released by Nokia.[186] The majority of old mobile phones are simply lying in drawers at home and very few old devices, about 4%, are being thrown into landfill and not recycled.[186]

All of Nokia's new models of chargers meet or exceed the Energy Star requirements.[187] Nokia aims to reduce its carbon dioxide emissions by at least 18 percent in 2010 from a baseline year of 2006 and cover 50 percent of its energy needs through renewable energy sources.[188] Greenpeace is challenging the company to use its influence at the political level as number 85 on the Fortune 500 to advocate for climate legislation and call for global greenhouse gas emissions to peak by 2015.[189]

Nokia is researching the use of recycled plastics in its products, which are currently used only in packaging but not yet in mobile phones.[190]

Since 2001, Nokia has provided eco declarations of all its products and since May 2010 provides Eco profiles for all its new products.[191] In an effort to further reduce their environmental impact in the future, Nokia released a new phone concept, Remade, in February 2008.[192] The phone has been constructed of solely recyclable materials.[192]

The outer part of the phone is made from recycled materials such as aluminium cans, plastic bottles, and used car tires.[193] The screen is constructed of recycled glass, and the hinges have been created from rubber tires. The interior of the phone is entirely constructed with refurbished phone parts, and there is a feature that encourages energy saving habits by reducing the backlight to the ideal level, which then allows the battery to last longer without frequent charges.

Research cooperation with universities

Nokia is actively exploring and engaging in open innovation through selective research collaborations with major universities and institutions by sharing resources and leveraging ideas. Major research collaboration is with Tampere University of Technology based in Finland. Current collaborations include:[194]

- Aalto University School of Science and Technology, Finland
- École Polytechnique Fédérale de Lausanne, Switzerland
- ETH Zurich, Switzerland
- Massachusetts Institute of Technology, United States
- Stanford University, United States
- Tampere University of Technology, Finland
- Tsinghua University, China
- University of California, Berkeley, United States
- University of Cambridge, United Kingdom
- University of Southern California, United States

Awards and recognition

The Brand Trust Report [195] published by Trust Research Advisory has ranked Nokia in the 1st position among the brands in India.

See also

Lists

- List of Nokia products
- List of acquisitions by Nokia

General

- Nokia Ovi Suite – Nokia's next generation phone suite software.
- Nokia PC Suite – A software package, slated to be replaced by Nokia Ovi Suite.
- Nokia Beta Labs – Nokia beta applications.
- Nokia Software Updater – Mobile device firmware updater.
- Symbian – An open source operating system for mobile devices.
- Maemo – Software and development platform and an operating system.
- MeeGo – Merger of Nokia's Maemo and Intel's Moblin projects.
- Qt – A cross-platform application development framework.
- Gnokii – A suite of programs for communicating with mobile phones.
- Nokia Pure - Nokia's current corporate font

Other

- Nokia head office – Nokia's headquarters.
- Nokia, Finland – A Finnish town.
- Nokian Tyres – A Finnish manufacturer of tires split from Nokia Corporation in 1988.

- Nokian Footwear – A Finnish manufacturer of boots split from Nokia Corporation in 1990.
- Nokia Arena, Tel Aviv
- Mobile phone

References

[1] http://www.nasdaqomxnordic.com/shares/shareinformation?Instrument=HEX24311

[2] http://www.nyse.com/about/listed/quickquote.html?ticker=nok

[3] http://www.boerse-frankfurt.de/EN/index.aspx?pageID=23&Query=NOA3

[4] "Annual Results 2010" (http://www.nokia.com/results/Nokia_results2010Q4e.pdf) (PDF). Nokia Corporation. 27 January 2011. . Retrieved 27 January 2011.

[5] http://www.nokia.com/

[6] "Nokia in brief (2007)" (http://www.nokia.com/NOKIA_COM_1/About_Nokia/Sidebars_new_concept/Nokia_in_brief/InBriefJuly08.pdf) (PDF). Nokia Corporation. March 2008. . Retrieved 14 May 2008.

[7] "Gartner Says Sales of Mobile Devices in Second Quarter of 2011 Grew 16.5 Percent Year-on-Year; Smartphone Sales Grew 74 Percent" (http://www.gartner.com/it/page.jsp?id=1764714) (Press release). Gartner. 11 August 2011. . Retrieved 29 September 2011.

[8] "Company" (http://www.nokiasiemensnetworks.com/global/AboutUs/Company/?languagecode=en). Nokia Siemens Networks. . Retrieved 14 July 2009.

[9] "Nokia to acquire NAVTEQ" (http://www.nokia.com/A4136001?newsid=1157198) (Press release). Nokia Corporation. 1 October 2007. . Retrieved 22 March 2009.

[10] "Nokia – FAQ" (http://www.nokia.com/about-nokia/company/faq). Nokia Corporation. . Retrieved 16 March 2009.

[11] Kapanen, Ari (24 July 2007). "Ulkomaalaiset valtaavat pörssiyhtiöitä" (http://www.taloussanomat.fi/porssi-ja-raha/2007/07/24/Ulkomaalaiset+valtaavat+pörssiyhtiöitä/20071 7658/103) (in Finnish). *Taloussanomat*. . Retrieved 14 May 2008.

[12] "2011 Ranking of the Top 100 Brands" (http://www.interbrand.com/en/best-global-brands/best-global-brands-2008/best-global-brands-2011.aspx). *Interbrand*. 15 November 2011. . Retrieved 15 November 2010.

[13] http://www.icon-net.com/medialib/file/eurobrand2011-BRAND-RANKING-Brand-CorporationsEurope2011.pdf

[14] "World's Most Admired Companies 2011" (http://money.cnn.com/magazines/fortune/mostadmired/2011/snapshots/6652.html). Fortune. 2011. . Retrieved 12 November 2011.

[15] "*Global 500* 2011" (http://money.cnn.com/magazines/fortune/global500/2011/full_list/101_200.html). Fortune. 2011. . Retrieved 12 November 2011.

[16] "40% profit fall for Nokia" (http://www.insideireland.ie/index.cfm/section/news/ext/nokia001/category/1084). *InsideIreland.ie*. 23 July 2010. . Retrieved 17 October 2010.

[17] "Nokia Q2 2011 net sales EUR 9.3 billion, non-IFRS EPS EUR 0.06 (reported EPS EUR -0.10)" (http://press.nokia.com/2011/07/21/nokia-q2-2011-net-sales-eur-9-3-billion-non-ifrs-eps-eur-0-06-reported-eps-eur-0-10/). Nokia. 21 July 2011. . Retrieved 21 July 2011.

[18] "Apple Vs. Nokia: The Global Smartphone Rivalry" (http://www.theglobalist.com/storyid.aspx?StoryId=8616). *The Globalist*. . Retrieved 17 October 2010.

[19] "Apple overtakes Nokia in smartphone volumes: Troubled Finnish group suffers further sharp sales fall" (http://www.ft.com/intl/cms/s/0/4d7fd1e2-b38e-11e0-b56c-00144feabdc0.html). Financial Times. 21 July 2011. . Retrieved 21 July 2011 October.

[20] Brian, Matt (21 July 2011). "Official: Apple sold more iPhones than Nokia's entire range of smartphones in Q2" (http://thenextweb.com/apple/2011/07/21/official-apple-sold-more-iphones-than-nokias-entire-range-of-smartphones-in-q2/). thenextweb.com. . Retrieved 21 July 2011 October.

[21] "Nokia's first Windows Phone 7 handset: Is it enough?" (http://www.bbc.co.uk/news/technology-15460569). BBC News. 26 October 2011. . Retrieved 27 October 2011.

[22] "Nokia – Towards Telecommunications" (http://www.nokia.com/NOKIA_COM_1/About_Nokia/Sidebars_new_concept/Broschures/TowardsTelecomms.pdf) (PDF). Nokia Corporation. August 2000. . Retrieved 5 June 2008.

[23] "Nokia – Nokia's first century – Story of Nokia" (http://www.nokia.com/about-nokia/company/story-of-nokia/nokias-first-century). Nokia Corporation. . Retrieved 16 March 2009.

[24] "Nokia – The birth of Nokia – Nokia's first century – Story of Nokia" (http://www.nokia.com/about-nokia/company/story-of-nokia/nokias-first-century/the-birth-of-nokia). Nokia Corporation. . Retrieved 16 March 2009.

[25] Helen, Tapio. "Idestam, Fredrik (1838–1916)" (http://www.kansallisbiografia.fi/english/?id=4296). Biographical Centre of the Finnish Literature Society. . Retrieved 22 March 2009.

[26] "Nokian Footwear: History" (http://www.nokianfootwear.fi/eng/our_story/). Nokian Footwear. . Retrieved 21 March 2009.

[27] Palo-oja, Ritva; Willberg, Leena (1998) (in Finnish). *Kumi – Kumin ja Suomen kumiteollisuuden historia*. Tampere, Finland: Tampere Museums. pp. 43–53. ISBN 9789516090651.

[28] "Finnish Cable Factory – Brief History" (http://web.archive.org/web/20070705233815/http://www.kaapelitehdas.fi/php/image.php?id=4856) (PDF). Kaapelitehdas.fi (http://www.kaapelitehdas.fi). Archived from the original (http://www.kaapelitehdas.fi/php/image.php?id=4856) on 5 July 2007. . Retrieved 16 March 2009.

[29] "Nokia – Verner Weckman – Nokia's first century – Story of Nokia" (http://www.nokia.com/about-nokia/company/story-of-nokia/nokias-first-century/verner-weckman). Nokia Corporation. . Retrieved 20 March 2009.

[30] "Nokia – The merger – Nokia's first century – Story of Nokia" (http://www.nokia.com/about-nokia/company/story-of-nokia/nokias-first-century/the-merger). Nokia Corporation. . Retrieved 16 March 2009.

[31] "Nokia – First electronic dept – Nokia's first century – Story of Nokia" (http://www.nokia.com/about-nokia/company/story-of-nokia/nokias-first-century/first-electronic-dept). Nokia Corporation. . Retrieved 16 March 2009.

[32] "Nokia – Jorma Ollila – Mobile revolution – Story of Nokia" (http://www.nokia.com/about-nokia/company/story-of-nokia/mobile-revolution/jorma-ollila). Nokia Corporation. . Retrieved 21 March 2009.

[33] "History in brief" (http://www.nokiantyres.com/history-in-brief). Nokian Tyres. . Retrieved 22 March 2009.

[34] Kaituri, Tommi (2000). "Automaattisten puhelinkeskusten historia" (http://www.cs.helsinki.fi/u/kerola/tkhist/k2000/alustukset/puhelinkeskukset/) (in Finnish). . Retrieved 21 March 2009.

[35] Palmberg, Christopher; Martikainen, Olli (23 May 2003). "Overcoming a Technological Discontinuity – The Case of the Finnish Telecom Industry and the GSM" (http://www.etla.fi/files/677_dp855.pdf) (PDF). The Research Institute of the Finnish Economy. . Retrieved 14 June 2009.

[36] "Puolustusvoimat: Kalustoesittely – Sanomalaitejärjestelmä" (http://www.mil.fi/maavoimat/kalustoesittely/index.dsp?level=81) (in Finnish). The Finnish Defence Forces. 15 June 2005. . Retrieved 14 May 2008.

[37] "The Finnish Defence Forces: Presentation of equipment: Message device" (http://www.mil.fi/maavoimat/kalustoesittely/00030_en.dsp). The Finnish Defence Forces. . Retrieved 14 June 2009.

[38] "Nokia 1100 phone offers reliable and affordable mobile communications for new growth markets" (http://press.nokia.com/PR/200308/915317_5.html) (Press release). Nokia Corporation. 27 August 2003. . Retrieved 26 May 2009.

[39] "Nokia – Mobira Cityman – The move to mobile – Story of Nokia" (http://www.nokia.com/about-nokia/company/story-of-nokia/the-move-to-mobile/mobira-cityman). Nokia Corporation. . Retrieved 14 May 2008.

[40] Juutilainen, Matti. "Siirtyvä tietoliikenne, luennot 7–8: Matkapuhelinverkot" (http://www.it.lut.fi/kurssit/06-07/Ti5312600/luentokalvot/luento07-08.pdf) (in Finnish) (PDF). Lappeenranta University of Technology. . Retrieved 22 March 2009.

[41] "Nokia – Mobile era begins – The move to mobile – Story of Nokia" (http://www.nokia.com/about-nokia/company/story-of-nokia/the-move-to-mobile/mobile-era-begins). Nokia Corporation. . Retrieved 20 March 2009.

[42] Karttunen, Anu (2 May 2003). "Tähdet syöksyvät, Benefon" (http://www.talouselama.fi/sijoittaminen/article165594.ece) (in Finnish). *Talouselämä* (Talentum Oyj). . Retrieved 28 July 2009.

[43] "Nokia´s Pioneering GSM Research and Development to be Awarded by Eduard Rhein Foundation" (http://press.nokia.com/PR/199710/776687_5.html) (Press release). Nokia Corporation. 17 October 1997. . Retrieved 22 March 2009.

[44] "Global Mobile Communication is 20 years old" (http://gsmworld.com/newsroom/press-releases/2070.htm) (Press release). GSM Association. 6 September 2007. . Retrieved 23 March 2009.

[45] "Happy 20th birthday, GSM" (http://news.zdnet.co.uk/leader/0,1000002982,39289154,00.htm). *ZDNet.co.uk* (CBS Interactive). 7 September 2007. . Retrieved 23 March 2009.

[46] "Nokia – First GSM call – The move to mobile – Story of Nokia" (http://www.nokia.com/about-nokia/company/story-of-nokia/the-move-to-mobile/first-gsm-call). Nokia Corporation. . Retrieved 20 March 2009.

[47] Smith, Tony (9 November 2007). "15 years ago: the first mass-produced GSM phone" (http://www.reghardware.co.uk/2007/11/09/ft_nokia_1011/). *Register Hardware*. Situation Publishing Ltd. . Retrieved 23 March 2009.

[48] "Nokia – Nokia Tune – Mobile revolution – Story of Nokia" (http://www.nokia.com/about-nokia/company/story-of-nokia/mobile-revolution/nokia-tune). Nokia Corporation. . Retrieved 23 March 2009.

[49] "3 Billion GSM Connections On The Mobile Planet – Reports The GSMA" (http://www.gsmworld.com/newsroom/press-releases/2008/1108.htm). GSM Association. 16 April 2008. . Retrieved 21 March 2009.

[50] "Nokia MikroMikko 1" (http://www.old-computers.com/museum/computer.asp?st=1&c=630). Old-Computers.com. . Retrieved 14 May 2008.

[51] "Net – Fujitsun asiakaslehti, Net-lehden historia: 1980-luku" (http://www.fujitsuservices.fi/historia/net/1980.htm) (in Finnish). Fujitsu Services Oy, Finland. . Retrieved 22 March 2009.

[52] "Historia: 1991–1999" (http://www.fujitsu.com/fi/about/history/1991/) (in Finnish). Fujitsu Services Oy, Finland. . Retrieved 22 March 2009.

[53] Hietanen, Juha (28 February 2000). "Closure of Fujitsu Siemens plant – a repeat of Renault Vilvoorde?" (http://www.eiro.eurofound.eu.int/2000/02/feature/fi0002136f.html). EIRO, European Industrial Relations Observatory on-line. . Retrieved 14 May 2008.

[54] Hietanen, Juha (28 February 2000). "Fujitsu Siemens tehdas suljetaan – toistuiko Renault Vilvoord?" (http://www.eurofound.europa.eu/eiro/2000/02/word/fi0002136ffi.doc) (in Finnish) (DOC). EIRO, European Industrial Relations Observatory on-line. . Retrieved 14 May 2008.

[55] "ViewSonic Corporation Acquires Nokia Display Products' Branded Business" (http://press.nokia.com/PR/200001/775025_5.html) (Press release). Nokia Corporation. 17 January 2000. . Retrieved 22 March 2009.

[56] "Nokia Booklet 3G brings all day mobility to the PC world" (http://www.nokia.com/press/press-releases/showpressrelease?newsid=1336683) (Press release). Nokia Corporation. 24 August 2009. . Retrieved 26 August 2009.

[57] Pietilä, Antti-Pekka (27 September 2000). "Kari Kairamon nousu ja tuho" (http://www.taloussanomat.fi/arkisto/2000/09/27/kari-kairamon-nousu-ja-tuho/20002643/12) (in Finnish). *Taloussanomat*. . Retrieved 21 March 2009.

[58] "Finland: How bad policies turned bad luck into a recession" (http://www.cepr.org/PRESS/EP29 finland.htm). Centre for Economic Policy Research. . Retrieved 5 April 2009.

[59] Häikiö, Martti; translated by Hackston, David (2001) (in Finnish). *Nokia Oyj:n historia 1–3 (A history of Nokia plc 1–3)* (http://www.finlit.fi/booksfromfinland/bff/102/nokia.htm). Helsinki: Edita. ISBN 951-37-3467-6. . Retrieved 21 March 2008.

[60] "Nokia – Leading the world – Mobile revolution – Story of Nokia" (http://www.nokia.com/about-nokia/company/story-of-nokia/mobile-revolution/leading-the-world). Nokia Corporation. . Retrieved 21 March 2009.

[61] Reinhardt, Andy (3 August 2006). "Nokia's Magnificent Mobile-Phone Manufacturing Machine" (http://www.businessweek.com/globalbiz/content/aug2006/gb20060803_618811.htm). *BusinessWeek Online Europe*. . Retrieved 21 March 2009.

[62] Professor Voomann, Thomas E.; Cordon, Carlos (1998). "Nokia Mobile Phones: Supply Line Management" (http://stuff.mit.edu/afs/athena/course/15/15.795/Nokia Supply Chain Case Study.pdf) (PDF). Lausanne, Switzerland: IMD – International Institute for Management Development. . Retrieved 21 March 2009.

[63] Ewing, Jack (30 July 2007). "Why Nokia Is Leaving Moto in the Dust" (http://www.businessweek.com/magazine/content/07_31/b4044050.htm). *BusinessWeek Online*. . Retrieved 21 March 2009.

[64] Lin, Porter; Khan, Raedeep; Piekute, Vaida; Luhtasela, Jussi; Fang, Debby (1 December 2005). "Supply Chain Management Case Nokia" (http://imba.nccu.edu.tw/OIP/EXchange/Docs/F04/mis/final/group6/SCM in Nokia - Written report-V1.0.pdf) (PDF). IMBA, College of Commerce, National Chengchi University. . Retrieved 21 March 2009.

[65] Virki, Tarmo (5 March 2007). "Nokia's cheap phone tops electronics chart" (http://www.reuters.com/article/technologyNews/idUSL0262945620070503). Reuters. . Retrieved 14 May 2008.

[66] "Nokia World 2007: Nokia outlines its vision of Internet evolution and commitment to environmental sustainability" (http://www.nokia.com/A4136001?newsid=1172937) (Press release). Nokia Corporation. 4 December 2007. . Retrieved 14 May 2008.

[67] "Nokia Productions and Spike Lee premiere the world's first social film" (http://www.nokia.com/press/press-releases/showpressrelease?newsid=1259528) (Press release). Nokia Corporation. 14 October 2008. . Retrieved 12 June 2009.

[68] "Nokia seizes social internet and amplifies music experience" (http://www.nokia.com/press/press-releases/showpressrelease?newsid=1338896) (Press release). Nokia Corporation. 2 September 2009. . Retrieved 12 October 2009.

[69] "Nokia 7705 Twist launched Stateside on Verizon (photo gallery)" (http://conversations.nokia.com/2009/09/10/nokia-7705-twist-launched-stateside-on-verizon-photo-gallery/). Nokia Corporation. 10 September 2009. . Retrieved 11 October 2009.

[70] "Hungarian and Finnish Prime Ministers Inaugurate Nokia's "Factory of the Future" in Komárom" (http://press.nokia.com/PR/200005/780293_5.html) (Press release). Nokia Corporation. 5 May 2000. . Retrieved 22 March 2009.

[71] "Nokia to set up a new mobile device factory in Romania" (http://www.nokia.com/A4136001?newsid=1114420) (Press release). Nokia Corporation. 26 March 2007. . Retrieved 14 May 2008.

[72] "Nokia to open cell phone plant near Cluj" (http://web.archive.org/web/20080507194303/http://www.boston.com/news/world/europe/articles/2007/03/22/nokia_to_open_cell_phone_plant_near_cluj/). Associated Press. Boston.com. 22 March 2007. Archived from the original (http://www.boston.com/news/world/europe/articles/2007/03/22/nokia_to_open_cell_phone_plant_near_cluj/) on 7 May 2008. . Retrieved 14 May 2008.

[73] "Nokia to build mobile phone plant in Romania" (http://www.hs.fi/english/article/Nokia+to+build+mobile+phone+plant+in+Romania/1135226144930). *Helsingin Sanomat*. 27 March 2007. . Retrieved 14 May 2008.

[74] "German Politicians Return Cell Phones Amid Nokia Boycott Calls" (http://www.dw-world.de/dw/article/0,2144,3076534,00.html). *Deutsche Welle*. 18 January 2008. . Retrieved 22 March 2009.

[75] "German State Demands €60 Million from Nokia" (http://www.spiegel.de/international/business/0,1518,540699,00.html). *Der Spiegel*. 11 March 2008. . Retrieved 22 March 2009.

[76] "Nokia Networks takes strong measures to reduce costs, improve profitability and strengthen leadership position" (http://press.nokia.com/PR/200304/898905_5.html) (Press release). Nokia Corporation. 10 April 2003. . Retrieved 14 May 2008.

[77] "Nokia Networks to shed 1,800 jobs worldwide; majority of impact felt in Finland" (http://www2.hs.fi/english/archive/news.asp?id=20030411IE6). *Helsingin Sanomat*. 11 April 2003. . Retrieved 14 May 2008.

[78] Leyden, John (10 April 2003). "Nokia Networks axes 1,800 staff" (http://www.theregister.co.uk/2003/04/10/nokia_networks_axes/). *The Register*. . Retrieved 14 May 2008.

[79] "Nokia's Law (transcription)" (http://www.yle.fi/mot/kj050117/englishscript.htm). YLE TV1, Mot. 17 January 2005. . Retrieved 14 May 2008.

[80] "Nokia and Sanyo proposed new company will not proceed" (http://www.nokia.com/A4136002?newsid=1059331) (Press release). Nokia Corporation. 26 June 2006. . Retrieved 14 May 2008.

[81] "Nokia decides not to go forward with Sanyo CDMA partnership and plans broad restructuring of its CDMA business" (http://www.nokia.com/A4136002?newsid=1059329) (Press release). Nokia Corporation. 22 June 2006. . Retrieved 14 May 2008.

[82] "Nokia and Sanyo Announce Intent to Form a Global CDMA Mobile Phones Business" (http://www.nokia.com/A4136002?newsid=1034612) (Press release). Nokia Corporation. 14 February 2006. . Retrieved 14 May 2008.

[83] "Shell appoints Jorma Ollila as new Chairman" (http://www.shell.com/home/content/media/news_and_library/press_releases/2005/pr_announcement_04082005.html) (Press release). Royal Dutch Shell. 4 August 2005. . Retrieved 22 March 2009.

[84] "Nokia moves forward with management succession plan" (http://www.nokia.com/A4136002?newsid=1004430) (Press release). Nokia Corporation. 1 August 2005. . Retrieved 22 March 2009.

[85] Repo, Eljas; Melender, Tommi (19 September 2005). "Changing the guard at Nokia – Olli-Pekka Kallasvuo takes the helm" (http://finland.fi/netcomm/news/showarticle.asp?intNWSAID=41296&LAN=ENG). *Ministry for Foreign Affairs of Finland*. Virtual Finland. . Retrieved 22 March 2009.

[86] Kallasvuo, Olli-Pekka; President and CEO (8 May 2008). "2008 Nokia Annual General Meeting (transcription)" (http://nds1.nokia.com/NOKIA_COM_1/Microsites/AGM_2008/pdf/OPK_AGM_2008_ENGLISH.pdf) (PDF). Helsinki Fair Centre, Amfi Hall: Nokia Corporation. . Retrieved 12 June 2009.

[87] " " (http://www.nokia.co.jp/about/release_081127.shtml) (in Japanese). – –. 27 November 2008. . Retrieved 5 December 2008.

[88] "Nokia completes acquisition of assets of Sega.com Inc." (http://www.nokia.com/A4136002?newsid=918198) (Press release). Nokia Corporation. 22 September 2003. . Retrieved 16 March 2009.

[89] "Nokia to extend leadership in enterprise mobility with acquisition of Intellisync" (http://www.nokia.com/A4136002?newsid=1021663) (Press release). Nokia Corporation. 16 November 2005. . Retrieved 22 March 2009.

[90] "Nokia completes acquisition of Intellisync" (http://www.nokia.com/A4136002?newsid=1034184) (Press release). Nokia Corporation. 10 February 2006. . Retrieved 22 March 2009.

[91] "Nokia and Siemens to merge their communications service provider businesses" (http://www.nokia.com/A4136002?newsid=1057716) (Press release). Nokia Corporation. 19 June 2006. . Retrieved 22 March 2009.

[92] "Nokia to acquire Loudeye and launch a comprehensive mobile music experience" (http://www.nokia.com/A4136002?newsid=1067845) (Press release). Nokia Corporation. 8 August 2006. . Retrieved 14 May 2008.

[93] "Nokia completes Loudeye acquisition" (http://www.nokia.com/A4136001?newsid=1081455) (Press release). Nokia Corporation. 16 October 2006. . Retrieved 14 May 2008.

[94] "Nokia acquires Twango to offer a comprehensive media sharing experience" (http://www.nokia.com/A4136001?newsid=1141417) (Press release). Nokia Corporation. 24 July 2007. . Retrieved 14 May 2008.

[95] "Nokia Acquires Twango – Frequently Asked Questions (FAQ)" (http://www.nokia.com/NOKIA_COM_1/Press/Materials/NokiaTwangoFAQ.pdf) (PDF). Nokia Corporation. . Retrieved 14 May 2008.

[96] "Nokia to acquire Enpocket to create a global mobile advertising leader" (http://www.nokia.com/A4136001?newsid=1153772) (Press release). Nokia Corporation. 17 September 2007. . Retrieved 14 May 2008.

[97] Niccolai, James (1 October 2007). "Nokia buys mapping service for $8.1 billion" (http://www.infoworld.com/article/07/10/01/Nokia-buys-mapping-service-for-8.1-billion_1.html). *IDG News Service* (InfoWorld). . Retrieved 14 May 2008.

[98] "Nokia completes its acquisition of NAVTEQ" (http://www.nokia.com/A4136001?newsid=1235107) (Press release). Nokia Corporation. 10 July 2008. . Retrieved 22 March 2009.

[99] "Nokia to acquire leading consumer email and instant messaging provider OZ Communications" (http://news.taume.com/World-Business/Business-Finance/Nokia-to-acquire-leading-consumer-email-and-instant-messaging-provider-OZ-Communications-6922). *Taume News*. 30 September 2008. . Retrieved 30 September 2008.

[100] "Nokia to acquire cellity" (http://www.nokia.com/press/press-releases/showpressrelease?newsid=1330831) (Press release). Nokia Corporation. 24 July 2009. . Retrieved 4 August 2009.

[101] "Nokia completes acquisition of cellity" (http://www.nokia.com/press/press-releases/showpressrelease?newsid=1332884) (Press release). Nokia Corporation. 5 August 2009. . Retrieved 6 August 2009.

[102] "Nokia has acquired Plum" (http://www.nokia.com/press/press-releases/showpressrelease?newsid=1340931) (Press release). Nokia Corporation. 11 September 2009. . Retrieved 28 January 2010.

[103] "Nokia Acquires Browser Firm Novarra" (http://techie-buzz.com/mobile-news/nokia-acquires-browser-firm-novarra.html). Techie-buzz. 28 March 2010. . Retrieved 29 March 2010.

[104] "Nokia Acquires MetaCarta" (http://www.informationweek.com/news/mobility/smart_phones/showArticle.jhtml?articleID=224202519). informationweek. 11 April 2010. . Retrieved 12 April 2010.

[105] "Nokia to cut 3500 jobs worldwide; to shut Romania factory" (http://www.moneycontrol.com/news/world-news/nokia-to-cut-3500-jobs-worldwide-to-shut-romania-factory_592235.html). 29 September 2011. .

[106] "Nokia to Cut 4,000 Jobs" (http://online.wsj.com/article/SB10001424052970204136404577210401816583074.html). February 8, 2012. .

[107] "Home of the Maemo community" (http://maemo.org/). maemo.org. . Retrieved 12 November 2011.

[108] Paul, Ryan (25 June 2010). "Nokia picks MeeGo Linux, not Symbian, for flagship phones" (http://arstechnica.com/open-source/news/2010/06/nokia-to-use-meego-linux-and-not-symbian-for-flagship-phones.ars). Ars Technica. . Retrieved 12 November 2011.

[109] Sherwood, James (13 August 2009). "Nokia exec denies Symbian Maemo swap claim" (http://www.reghardware.co.uk/2009/08/13/nokia_denies_maemo/). reghardware. . Retrieved 12 November 2011.

[110] "Nokia announces strategic partnership with Microsoft, will use WP7 as primary OS" (http://www.techit.in/2011/02/nokia-announces-strategic-partnership-with-microsoft-will-use-wp7-as-primary-os/). TechIt.in. .

[111] "Missed the historic Nokia+Microsoft event today? See it here!" (http://www.techit.in/2011/02/missed-the-historic-nokiamicrosoft-event-today-see-it-here/). TechIt.in. .

[112] "Nokia and Microsoft form partnership" (http://www.bbc.co.uk/news/business-12427680). BBC. 11 February 2011. . Retrieved 12 February 2011.

[113] http://www.tomshw.it/cont/news/nokia-crolla-in-borsa-a-vincere-e-microsoft/29696/3.html

[114] "RIP: Symbian" (http://www.engadget.com/2011/02/11/rip-symbian/). Engadget. .
[115] "Capitulation: Nokia adopts Windows Phone 7" (http://arstechnica.com/gadgets/news/2011/02/nokia-adopts-windows-phone-7-as-primary-platform.ars). ArsTechnica. 11 February 2011. . Retrieved 12 February 2011.
[116] "Nokia will be able to customize 'everything' in Windows Phone 7, but likely won't" (http://www.engadget.com/2011/02/11/nokia-will-be-able-to-customize-everything-in-windows-phone-7). Engadget. .
[117] ben-Aaron, Diana (11 February 2011). "Nokia Falls Most Since July 2009 After Microsoft Deal" (http://www.bloomberg.com/news/2011-02-11/nokia-joins-forces-with-microsoft-to-challenge-dominance-of-apple-google.html). Bloomberg. .
[118] Ward, Andrew (21 July 2011). "Apple overtakes Nokia in smartphone stakes" (http://www.ft.com/cms/s/0/4d7fd1e2-b38e-11e0-b56c-00144feabdc0.html#axzz1SlVal4Cs). *Financial Times*. . Retrieved 21 July 2011.
[119] Fried, Ina (9 August 2011). "Nokia to Exit Symbian, Low-End Phone Businesses in North America" (http://allthingsd.com/20110809/exclusive-nokia-to-exit-symbian-low-end-phone-businesses-in-north-america/). All Things Digital. . Retrieved 9 August 2011.
[120] "Structure" (http://www.nokia.com/about-nokia/company/structure). Nokia Corporation. 1 October 2009. . Retrieved 28 December 2009.
[121] "Nokia Siemens Networks starts operations and assumes a leading position in the communications industry" (http://www.nokia.com/A4136002?newsid=1116423) (Press release). Nokia Corporation. 2 April 2007. . Retrieved 7 April 2009.
[122] "Nokia's 25 percent profit jump falls short of expectations" (http://www.usatoday.com/money/economy/2008-04-17-173945271_x.htm). Associated Press. USA Today. 17 April 2008. . Retrieved 14 May 2008.
[123] "Nokia, LG lose while ZTE, Apple gain Q4 2010 market share" (http://www.mobileburn.com/news.jsp?Id=12703). mobileburn.com. 28 January 2011. .
[124] "The Wave of the Future" (http://www.underconsideration.com/brandnew/archives/the_wave_of_the_future.php). *Brand New: Opinions on Corporate and Brand Identity Work*. UnderConsideration LLC. 25 March 2007. . Retrieved 14 May 2008.
[125] "Reviews – 2007 – Nokia Siemens Networks" (http://www.identityworks.com/reviews/2007/Nokia_Siemens.htm). *Identityworks*. 2007. . Retrieved 14 May 2008.
[126] "Facts about Nokia Siemens Networks" (http://www.nokiasiemensnetworks.com/NR/rdonlyres/B905B5DD-63F3-4D39-8CF3-B78AC915462E/0/_Factsheet_March_09_final.pdf) (PDF). Nokia Siemens Networks. March 2009. . Retrieved 7 April 2009.
[127] "Trovicor Monitoring Center" (http://www.trovicor.com/de/business-section-german/monitoring-center.html) (http). Trovicor. 2011. . Retrieved 23 August 2011.
[128] "Torture in Bahrain Becomes Routine With Help From Nokia Siemens" (http://www.bloomberg.com/news/print/2011-08-22/torture-in-bahrain-becomes-routine-with-help-from-nokia-siemens-networking.html). *Bloomberg*. 22 August 2011. . Retrieved 23 August 2011.
[129] "Group Executive Board" (http://www.nokia.com/A4126335). Nokia Corporation. April 2007. . Retrieved 14 May 2008.
[130] "Board of Directors" (http://www.nokia.com/A4126350). Nokia Corporation. April 2007. . Retrieved 14 May 2008.
[131] "Audit Committee Charter at Nokia" (http://www.nokia.com/NOKIA_COM_1/About_Nokia/Sidebars_new_concept/Board_charters/audit_charter.pdf) (PDF). Nokia Corporation. 2007. . Retrieved 14 May 2008.
[132] "Personnel Committee Charter at Nokia" (http://www.nokia.com/NOKIA_COM_1/About_Nokia/Sidebars_new_concept/Board_charters/personnel_charter_2007.pdf) (PDF). Nokia Corporation. 2007. . Retrieved 14 May 2008.
[133] "Corporate Governance and Nomination Committee Charter at Nokia" (http://www.nokia.com/NOKIA_COM_1/About_Nokia/Sidebars_new_concept/Board_charters/CG_Charter_2008_Final_20080123.pdf) (PDF). Nokia Corporation. 2008. . Retrieved 14 May 2008.
[134] "Committees of the Board" (http://www.nokia.com/link?cid=EDITORIAL_4207). Nokia Corporation. May 2007. . Retrieved 14 May 2008.
[135] Virkkunen, Johannes (29 September 2006). "New Finnish Companies Act designed to increase Finland's competitiveness" (http://www.lmr.fi/publications/companies_act_290906.pdf) (PDF). *LMR Attorneys Ltd. (Luostarinen Mettälä Räikkönen)*. . Retrieved 14 May 2008.
[136] "Articles of Association" (http://www.nokia.com/NOKIA_COM_1/About_Nokia/Company/Corporate_Governance/Articles_of_Association/Nokia_Articles_of_Association_10052007.pdf) (PDF). Nokia Corporation. 10 May 2007. . Retrieved 14 May 2008.
[137] "Corporate Governance Guidelines at Nokia" (http://www.nokia.com/NOKIA_COM_1/About_Nokia/Sidebars_new_concept/Board_charters/corporate_governance_guideline_sep06.pdf) (PDF). Nokia Corporation. 2006. . Retrieved 14 May 2008.
[138] "Suomalaisten yritysten ylin johto" (http://www.kolumbus.fi/taglarsson/dokumentit/yritys.htm) (in Finnish). . Retrieved 20 March 2009.
[139] "Nokia Research Center" (http://www.nokia.com/NOKIA_COM_1/Press/twwln/press_kit/Nokia_Research_Center_Press_Backgrounder_October_2007.pdf) (PDF). Nokia Corporation. October 2007. . Retrieved 14 May 2008.
[140] "About NRC – Nokia Research Center" (http://research.nokia.com/aboutus/index.html). Nokia Corporation. . Retrieved 17 March 2009.
[141] "NRC Locations – Nokia Research Center" (http://research.nokia.com/locations/index.html). Nokia Corporation. . Retrieved 17 March 2009.
[142] "INdT – Instituto Nokia de Tecnologia" (http://www.indt.org.br/). Nokia Corporation. . Retrieved 17 March 2009.
[143] "Production units" (http://www.nokia.com/A4149133). Nokia Corporation. June 2008. . Retrieved 14 May 2008.
[144] "Nokia despre sechestrul ANAF: Colaborăm pentru a ne asigura că situaţia e soluţionată satisfăcător" (http://www.mediafax.ro/economic/nokia-despre-sechestrul-anaf-colaboram-pentru-a-ne-asigura-ca-situatia-e-solutionata-satisfacator-8963720) (Press release).

mediafax.ro. 12 November 2011. . Retrieved 12 November 2011.
[145] Kapanen, Ari (24 July 2007). "Ulkomaalaiset valtaavat pörssiyhtiöitä" (http://www.taloussanomat.fi/porssi-ja-raha/2007/07/24/Ulkomaalaiset+valtaavat+pörssiyhtiöitä/200717658/103) (in Finnish). *Taloussanomat*. . Retrieved 14 May 2008.
[146] Ali-Yrkkö, Jyrki (2001). "The role of Nokia in the Finnish Economy" (http://www.etla.fi/files/940_FES_01_1_nokia.pdf) (PDF). ETLA (The Research Institute of the Finnish Economy). . Retrieved 21 March 2009.
[147] Ali-Yrkkö, Jyrki (2010). "NOKIA AND FINLAND IN A SEA OF CHANGE" (http://www.etla.fi/files/2585_nokia_kirja_8_v2_kansineen.pdf). *ETLA – Research Institute of the Finnish Economy*. . Retrieved 12 November 2011.
[148] Guardian newspaper: Nokia cuts 4,000 jobs and moves smartphone manufacturing to Asia, 9 February 2012 (http://www.guardian.co.uk/technology/2012/feb/08/nokia-cuts-4000-manufacturing-jobs)
[149] "HS Archives" (http://www.hs.fi/arkisto/haku?pageNumber=1&order=FIFO&advancedSearch=&free=Connecting+People+Ove+Strandberg&date=year2003&depa=Kaikki+osastot&fromDay=0&fromMonth=0&fromYear=0&toDay=0&toMonth=0&toYear=0) (in Finnish). Helsingin Sanomat. 1 June 2003. . Retrieved 14 May 2008.
[150] "NOKIA | Connecting Pople 1992 Vector Logo (AI EPS)" (http://www.hdicon.com/vector-logos/nokia-connecting-pople-1992/). *HDicon.com*. . Retrieved 17 October 2010.
[151] Pitkänen, Juhani (Nokia's Art Director) (3 September 2007). "Nokia Strategic Marketing, Brand Identity" (http://www.nokia.com/A4126575). Nokia Corporation. . Retrieved 14 May 2008.
[152] "NOKIA | Connecting Pople new Vector Logo (AI EPS)" (http://www.hdicon.com/vector-logos/nokia-connecting-pople-new/). *HDicon.com*. . Retrieved 17 October 2010.
[153] "Erik Spiekermann – Furniture, Designs & Home Decor" (http://www.dwr.com/category/designers/r-t/erik-spiekermann.do). Design Within Reach. . Retrieved 7 January 2010.
[154] "Nokia on the ropes as analysts slash targets" (http://beststockpicks.biz/earnings/nokia-on-the-ropes-as-analysts-slash-targets-2.html/). Beststockpicks.biz. . Retrieved 12 November 2011.
[155] http://media.corporate-ir.net/media_files/IROL/10/107224/Nokia_results2011Q2e.pdf
[156] "Nokia Way and values" (http://www.nokia.com/A4126303). Nokia Corporation. . Retrieved 14 May 2008.
[157] "dotMobi Investors" (http://web.archive.org/web/20080507214157/http://mtld.mobi/company/about/investors). dotMobi. Archived from the original (http://mtld.mobi/company/about/investors) on 7 May 2008. . Retrieved 14 May 2008.
[158] Haumont, Serge; Siren, Ritva. "dotMobi, a Key Enabler for the Mobile Internet" (http://research.nokia.com/files/Haumont-dotMobi.pdf) (PDF). *Nokia Research Center*. Nokia Corporation. . Retrieved 14 May 2008.
[159] http://nokia.mobi
[160] "Nokia Ad Business" (http://www.adservice.nokia.com/faq.jsp#12). Nokia Corporation. . Retrieved 14 May 2008.
[161] http://www.adservice.nokia.com/
[162] Reardon, Marguerite (6 March 2007). "Nokia introduces mobile ad services" (http://news.com.com/2100-1039_3-6164800.html). *CNET News.com*. . Retrieved 14 May 2008.
[163] "Meet Ovi, the door to Nokia's Internet services" (http://www.nokia.com/A4136001?newsid=1149749) (Press release). Nokia Corporation. 29 August 2007. . Retrieved 7 April 2009.
[164] Niccolai, James (4 December 2007). "Nokia Lays Plan for More Internet Services" (http://www.nytimes.com/idg/IDG_002570DE00740E18002573A70046F2EF.html?ref=technology). *IDG News Service* (New York Times). . Retrieved 14 May 2008.
[165] "Ovi by Nokia" (http://www.nokia.com/NOKIA_COM_1/Press/Materials/White_Papers/pdf_files/backgrounders2008/Backgrounder_Ovi_by_Nokia.pdf) (PDF). Nokia Corporation. . Retrieved 7 April 2009.
[166] "Ovi Store opens for business" (http://www.nokia.com/press/press-releases/showpressrelease?newsid=1317441) (Press release). Nokia Corporation. 26 May 2009. . Retrieved 12 June 2009.
[167] Virki, Tarmo (18 March 2009). "Nokia to shutter its "Mosh" success story" (http://www.reuters.com/article/technologyNews/idUSTRE52H6AI20090318?pageNumber=1&virtualBrandChannel=0). *Reuters*. . Retrieved 14 July 2009.
[168] "Nokia launches new online magazine" (http://www.newstatesman.com/magazines/2010/03/online-magazine-nokia-ovi). NewStatesman. 23 March 2010. . Retrieved 29 March 2010.
[169] "Nokia – My Nokia" (http://europe.nokia.com/my-nokia). Nokia Corporation. . Retrieved 10 January 2012.
[170] "Nokia Retreats from Music Service" (http://www.yle.fi/uutiset/news/2011/01/nokia_retreats_from_music_service_2294706.html). YLE. 18 January 2011. . Retrieved 18 January 2011.
[171] Fields, Davis (17 December 2008). "Nokia Email service graduates as part of Nokia Messaging" (http://betalabs.nokia.com/blog/2008/12/17/nokia-email-service-graduates-as-part-of-nokia-messaging/). *Nokia Beta Labs*. Nokia Corporation. . Retrieved 16 March 2009.
[172] "Nokia Messaging: FAQ" (http://email.nokia.com/account/faq.action?change_locale=en). Nokia Corporation. . Retrieved 12 June 2009.
[173] Cellan-Jones, Rory (22 June 2009). "Hi-tech helps Iranian monitoring" (http://news.bbc.co.uk/1/hi/technology/8112550.stm). *BBC News*. . Retrieved 14 July 2009.
[174] Rhoads, Christopher; Chao, Loretta (22 June 2009). "Iran's Web Spying Aided By Western Technology" (http://online.wsj.com/article/SB124562668777335653.html#mod). *The Wall Street Journal* (Dow Jones & Company, Inc.): pp. A1. . Retrieved 14 July 2009.
[175] "Provision of Lawful Intercept capability in Iran" (http://www.nokiasiemensnetworks.com/global/Press/Press+releases/news-archive/Provision+of+Lawful+Intercept+capability+in+Iran.htm) (Press release). Nokia Siemens Networks. 22 June 2009. . Retrieved 14 July 2009.

[176] Kamali Dehghan, Saeed (14 July 2009). "Iranian consumers boycott Nokia for 'collaboration'" (http://www.guardian.co.uk/world/2009/jul/14/nokia-boycott-iran-election-protests). *The Guardian* (London: Guardian News and Media Limited). . Retrieved 27 July 2009.
[177] Ozimek, John (6 March 2009). "'Lex Nokia' company snoop law passes in Finland" (http://www.theregister.co.uk/2009/03/06/finland_nokia_snooping/). *The Register*. . Retrieved 27 July 2009.
[178] "Nokia Denies Threat to Leave Finland" (http://www.cellular-news.com/story/35783.php). *cellular-news*. 1 February 2009. . Retrieved 27 July 2009.
[179] Virki, Tarmo (18 January 2010). "SCENARIOS-What lies ahead in Nokia vs Apple legal battle" (http://www.reuters.com/article/idUSLDE60H05R20100118?type=marketsNews). *Reuters*. . Retrieved 25 January 2010.
[180] "The war of the Smartphones: Nokia's new patent suit against Apple" (http://pda-phone-reviews.in/latest-news/nokias-new-patent-suit-against-apple/). *Snartphone Reviews*. 6 January 2010. . Retrieved 25 January 2010.
[181] "Nokia's Patent Settlement With Apple Won't Help Much" (http://www.informationweek.com/news/personal-tech/smart-phones/230600172). 14 Jun 2011. . Retrieved 29 June 2011.
[182] Smith, Catharine (14 Jun 2011). "Apple Settles With Nokia In Patent Lawsuit" (http://www.huffingtonpost.com/2011/06/14/apple-nokia-patent-lawsuit-settlement_n_876499.html). *Huffington Post*. . Retrieved 29 June 2011.
[183] ben-Aaron, Diana; Pohjanpalo, Kati (14 Jun 2011). "Nokia Wins Apple Patent-License Deal Cash, Settles Lawsuits" (http://www.bloomberg.com/news/2011-06-14/nokia-apple-payments-to-nokia-settle-all-litigation.html). *Bloomberg*. . Retrieved 29 June 2011.
[184] "Guide to Greener Electronics – Greenpeace International" (http://www.greenpeace.org/international/en/campaigns/climate-change/cool-it/Guide-to-Greener-Electronics/). Greenpeace International. . Retrieved 14 November 2011.
[185] "Nokia – Where and how to recycle – Recycling – Environment" (http://www.nokia.com/environment/recycling/where-and-how-to-recycle). Nokia. . Retrieved 12 August 2010.
[186] "Global consumer survey reveals that majority of old mobile phones are lying in drawers at home and not being recycled" (http://www.nokia.com/press/press-releases/showpressrelease?newsid=1234291) (Press release). Nokia Corporation. 8 July 2008. . Retrieved 27 July 2009.
[187] "Nokia – Energy efficiency – Devices and services – Environment" (http://www.nokia.com/environment/we-energise/nokia-and-energy-efficiency). Nokia. . Retrieved 12 August 2010.
[188] "Nokia – Energy saving targets – Environmental strategy – Strategy and reports – Environment" (http://www.nokia.com/environment/strategy-and-reports/environmental-strategy/energy-saving-targets). Nokia. . Retrieved 12 August 2010.
[189] "Nokia" (http://www.greenpeace.org/international/en/campaigns/toxics/electronics/Guide-to-Greener-Electronics/companies/Nokia/). Greenpeace International. . Retrieved 12 August 2010.
[190] "Materials and substances" (http://www.nokia.com/environment/we-create/materials-and-substances). Nokia Corporation. . Retrieved 12 August 2010.
[191] "Eco declarations" (http://www.nokia.com/environment/we-create/devices-and-accessories/eco-declarations). Nokia Corporation. . Retrieved 27 July 2009.
[192] Rubio, Jenalyn (12 April 2008). "Tech Goes Greener" (http://www.pcworld.com/article/id,144482-c,recycling/article.html). *Computerworld Philippines* (PC World). . Retrieved 14 May 2008.
[193] "Nokia Remade Concept Phone goes Green" (http://www.mobiletor.com/2008/04/09/nokia-remade-concept-phone-goes-green/). Mobiletor. 9 April 2008. . Retrieved 14 May 2008.
[194] "Open Innovation – Nokia Research Center" (http://research.nokia.com/openinnovation). Nokia Corporation. . Retrieved 1 April 2009.
[195] "Nokia is India's most trusted brand : Brand Trust" (http://profit.ndtv.com/news/show/nokia-is-india-s-most-trusted-brand-brand-trust-136783). profit.ndtv.com. 19 January 2011. . Retrieved 21 November 2011.

Further reading

Title	Author	Publisher	Year	Length	ISBN
Winning Across Global Markets: How Nokia Creates Strategic Advantage in a Fast-Changing World	Dan Steinbock	Jossey-Bass / Wiley	May 2010	304 pp	ISBN 9780470339664
Nokia: The Inside Story	Martti Häikiö	FT / Prentice Hall	October 2002	256 pp	ISBN 0-273-65983-9
Work Goes Mobile: Nokia's Lessons from the Leading Edge	Michael Lattanzi, Antti Korhonen, Vishy Gopalakrishnan	John Wiley & Sons	January 2006	212 pp	ISBN 0-470-02752-5
Mobile Usability: How Nokia Changed the Face of the Mobile Phone	Christian Lindholm, Turkka Keinonen, Harri Kiljander	McGraw-Hill Companies	June 2003	301 pp	ISBN 0-07-138514-2

Business The Nokia Way: Secrets of the World's Fastest Moving Company	Trevor Merriden	John Wiley & Sons	February 2001	168 pp	ISBN 1-84112-104-5
The Nokia Revolution: The Story of an Extraordinary Company That Transformed an Industry	Dan Steinbock	AMACOM Books	April 2001	375 pp	ISBN 0-8144-0636-X

External links

- Official Nokia international website (http://www.nokia.com/)

Nokia_Eseries

Manufacturer	Nokia
Series	Eseries
Availability by country	2005—present
Operating system	Symbian
Rear camera	1.3 MP—8.0 MP
Front camera	VGA
Ringtones & notifications	Nokia tune
Development status	Active since 2005

The **Nokia Eseries** consists of business-oriented smartphones, with emphasis on enhanced connectivity and support for corporate e-mail push services. All devices have advanced office features. Phones equipped with Wireless LAN also provide a VoIP client (SIP Protocol).

Announcements

- On October 12, 2005, mobile phone manufacturer Nokia announced what the company refers to as the Eseries, consisting of the three mobile phones, the Nokia E60, Nokia E61 and Nokia E70.[1]
- On May 18, 2006, Nokia announced the addition of the E50 to the series, which it refers to as a "business device" rather than a "smartphone".[2]
- On February 12, 2007, Nokia announced the addition of three new devices to the series; E61i, E65 and E90.[3]
- On April 11, 2008, Nokia Australia has advised that the E61i will be discontinued in May 2008 and be replaced by a more featured but smaller E71.

List of Eseries phones

Nokia E72 is officially announced on June 15, 2009 – 9:48 am.

Nokia E63

Nokia E71

Nokia E7

Connectivity and synchronization

Nokia provides the Nokia PC Suite software that allows synchronization and transfer of data, calendars, contacts, and more between most Nokia phone and a Windows system via USB, Bluetooth or Infrared,the latter two being mostly used with portable systems like laptops,due to lack of Bluetooth or Infrared support in most PC's,specially in legacy systems.

Mac users can access the data on an E-series phone using a bluetooth connection, e.g. the Bluetooth File Exchange application native to OS X. Synchronizing addresses and calendars is done using iSync: some Nokia phones are supported natively by Mac OS X, some require that the user installs an iSync plugin from Nokia.[4]

Nokia also provides Nokia Multimedia Transfer[5] for managing user content on the device. NMT will synchronize music, videos and photos from selected iTunes playlists and iPhoto albums into the phone, transcoding and reducing the file size if necessary. It also presents the device as a digital camera in iPhoto for easy importing of photos and videos.

See also

- Nokia Cseries
- Nokia Nseries
- Nokia Xseries
- List of Nokia products

References

[1] "New Nokia family of devices targeted at the business world" (http://press.nokia.com/PR/200510/1015195_5.html) (Press release). Nokia. 2005-10-12. . Retrieved 2006-05-28.

[2] "The smallest of Nokia Eseries, the Nokia E50 business device for mobile professionals" (http://press.nokia.com/PR/200605/1051798_5.html) (Press release). Nokia. 2006-05-18. . Retrieved 2006-05-28.

[3] "Nokia unleashes second wave of Nokia Eseries business devices" (http://www.nokia.com/A4136001?newsid=1104232) (Press release). Nokia. 2007-02-12. . Retrieved 2007-02-14.

[4] Nokia's support page for Mac OS X (http://nokia.com/mac)

[5] Apple's download page for Nokia Multimedia Transfer (http://www.apple.com/downloads/macosx/drivers/nokiamultimediatransfer.html)

External links

- Nokia Eseries (Product Portal) (http://www.nokiaforbusiness.com/nfb/find_a_product/browse_mobile_devices.html)
- Detail Information for all Nokia devices (http://www.forum.nokia.com/devices/)
- "Nokia E62 (finally) hits Cingular" (http://www.engadgetmobile.com/2006/09/12/nokia-e62-finally-hits-cingular/). Engadget Mobile. 2006-09-12. Retrieved 2006-09-12.
- "First Look: Nokia E61" (http://www.ohgizmo.com/2006/06/08/ohgizmo-first-look-nokia-e61/). OhGizmo!. 2006-06-08. Retrieved 2006-06-14.
- "E60 photos" (http://www.flickr.com/photos/tags/nokiae60/). Flickr. 2006-06-08. Retrieved 2006-06-12.
- "Nokia's E60, E61, and E70 E-series mobile business phones" (http://www.engadget.com/2005/10/12/nokias-e60-e61-and-e70-e-series-mobile-business-phones/). Engadget. 2005-10-12. Retrieved 2006-05-28.
- "Nokia Eseries" (http://www.phonescoop.com/articles/digital_2005/index.php?p=e7). phonescoop.com. 2005-10-14. Retrieved 2006-05-28.
- "VoIP on the Nokia E-Series" (http://www.aql.com/telecoms/site/voip-on-nokia.php). aql.com. 2006-10-19. Retrieved 2006-10-19.
- "How to set up VoIP with Asterisk on the E61 and E70" (http://cook.dannemann.org.uk/howto/e61sip/). Christian Dannemann. 2006-12-11. Retrieved 2006-12-11.

Multi-band

In telecommunications, the terms **multi-band**, **dual-band**, **tri-band**, **quad-band** and **penta-band** refer to a device (especially a mobile phone) supporting multiple radio frequency bands. All devices which have more than one channel use multiple frequencies; a band however is a group of frequencies containing many channels. Multiple bands in mobile devices support roaming between different regions where different standards are used for mobile telephone services. Where the bands are widely separated in frequency, parallel transmit and receive signal path circuits must be provided, which increases the cost, complexity and power demand of multi-band devices.

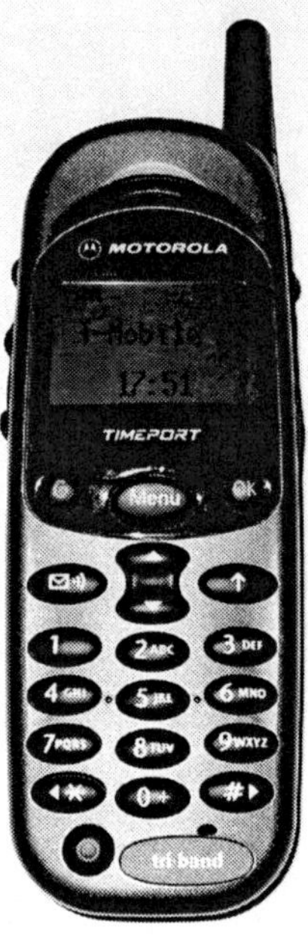

Motorola Timeport, the first tri-band mobile phone (1999)

Quad-Band

The term quad-band describes four ("quad") frequency bands: 850 and 1900 MHz, mostly used in Canada and the United States, and the common 900 and 1800 MHz bands.

See also

- List of UMTS networks around the world, and the frequencies and data rates they support
- GSM frequency bands, as defined by the standards bodies
- UMTS frequency bands, as defined by the standards bodies

Intellisync

The **Intellisync Corporation** is a provider of Data and PIM Synchronization software for mobile devices, such as Cell Phones and PDAs. The company is currently a part of Nokia, after it was acquired in 2006.

History

The company was formed after the San Jose, California, based Pumatech acquired the Alpharetta, Georgia based Synchrologic in late 2003 and renamed itself in 2004. The company has won several awards for its mobile software and data synchronization software. On January 31, 2006, stockholder approval was secured for Intellisync to be acquired by Nokia. On February 10, 2006, Nokia formally completed its acquisition of Intellisync.[1] Nokia has announced that the IntelliSync Desktop product has been discontinued and the last date available to order the product is July 19, 2008. Product support will be provided thru July 19, 2010 if eligible for support.Nokia Intellisync Mobile Suite remains a key product of Nokia Email strategy.

On September 29, 2008 Nokia announced that it plans to cease developing or marketing its own behind-the-firewall business mobility solutions (Intellisync Mobility Product Suite). The appropriate technologies and expertise will be reallocated to Nokia's new consumer push e-mail service. Going forward, Nokia plans to form its enterprise solutions offering by combining Nokia devices and applications with software solutions from industry leading enterprise vendors such as Microsoft, IBM, Cisco and others.

Locations

After the Nokia acquisition, the company headquarters in San Jose was closed and all employees moved to the Nokia office in Mountain View. The company maintains development offices in Alpharetta, Georgia, Bulgaria, New Delhi, Tokyo and Cluj-Napoca.

Products and Devices

Intellisync's flagship product, the Intellisync Mobile Suite, is a suite of software designed for corporations and wireless carriers. The Suite includes four modules, that can be installed independently or in any combination: Wireless Email, Systems/Device Management, File Sync and Application Sync. There is also a specific application reverse proxy sold as Intellisync Secure Gateway. This allows a more secure firewall configuration.

The intent is to provide database synchronization with the company's Application Sync product; Email and PIM synchronization with Wireless Email, and mobile device management and static file distribution with Device Management/File Sync. These products support synchronization with a corporation's Microsoft Exchange Server, Domino mail servers or Novell GroupWise, as well as POP and IMAP mail support.

The company originally only provided software for Windows-based computers, Palm Devices, Handheld PCs, and Pocket PCs, but has recently expanded into supporting Symbian, BREW, and other mobile devices. While the company originally marketed its product to large businesses (such as Boeing, Nintendo and the United States Military), it recently began rebranding its product to be distributed by wireless carriers as a revenue enhancing service to individual consumers. Wireless providers that have partnered with Intellisync include Verizon and Eurotel.

Yahoo is one of the internet providers that uses Intellisync products. AutoSync Yahoo is a product written by Intellisync to sync Yahoo contacts, notes, calendars and tasks with Outlook Express and MS Outlook.

Its primary competitor is Research in Motion (RIM).

Acquisitions

- July 1998, SoftMagic;
- October 1999, ProxiNet;
- February 2000, NetMind;
- July 2000, Dry Creek Software;
- October 2000; The Windward Group.

References

[1] http://ncomprod.nokia.com/about-nokia/financials/acquisitions

External links

- http://www.intellisync.com
- http://www.nokiaforbusiness.com

Smartphone

A **smartphone** is a mobile phone built on a mobile computing platform, with more advanced computing ability and connectivity than a feature phone.[1] [2] [3] The first smartphones were devices that mainly combined the functions of a personal digital assistant (PDA) and a mobile phone or camera phone. Today's models also serve to combine the functions of portable media players, low-end compact digital cameras, pocket video cameras, and GPS navigation units. Modern smartphones typically also include high-resolution touchscreens, web browsers that can access and properly display standard web pages rather than just mobile-optimized sites, and high-speed data access via Wi-Fi and mobile broadband.

Modern smartphones.

The most common mobile operating systems (OS) used by modern smartphones include Apple's iOS, Google's Android, Microsoft's Windows Phone, Nokia's Symbian, RIM's BlackBerry OS, and embedded Linux distributions such as Maemo and MeeGo. Such operating systems can be installed on many different phone models, and typically each device can receive multiple OS software updates over its lifetime.

The distinction between smartphones and feature phones can be vague and there is no official definition for what constitutes the difference between them. One of the most significant differences is that the advanced application programming interfaces (APIs) on smartphones for running third-party applications[4] can allow those applications to have better integration with the phone's OS and hardware than is typical with feature phones. In comparison, feature phones more commonly run on proprietary firmware, with third-party software support through platforms such as Java ME or BREW.[1] An additional complication in distinguishing between smartphones and feature phones is that over time the capabilities of new models of feature phones can increase to exceed those of phones that had been promoted as smartphones in the past.

History

Early years

IBM Simon (introduced 1992) shown in the charging station

The first smartphone was the IBM Simon; it was designed in 1992 and shown as a concept product[5] that year at COMDEX, the computer industry trade show held in Las Vegas, Nevada. It was released to the public in 1993 and sold by BellSouth. Besides being a mobile phone, it also contained a calendar, address book, world clock, calculator, note pad, e-mail client, the ability to send and receive faxes, and games. It had no physical buttons, instead customers used a touchscreen to select telephone numbers with a finger or create faxes and memos with an optional stylus. Text was entered with a unique on-screen "predictive" keyboard. By today's standards, the Simon would be a fairly low-end product, lacking a camera and the ability to download third-party applications. However, its feature set at the time was highly advanced.

The Nokia Communicator line was the first of Nokia's smartphones starting with the Nokia 9000, released in 1996. This distinctive palmtop computer style smartphone was the result of a collaborative effort of an early successful and costly personal digital assistant (PDA) by Hewlett-Packard combined with Nokia's best-selling phone around that time, and early prototype models had the two devices fixed via a hinge. The Communicators are characterized by a clamshell design, with a feature phone display, keyboard and user interface on top of the phone, and a physical QWERTY keyboard, high-resolution display of at least 640×200 pixels and PDA user interface under the flip-top. The software was based on the GEOS V3.0 operating system, featuring email communication and text-based web browsing. In 1998, it was followed by Nokia 9110, and in 2000 by Nokia 9110i, with improved web browsing capability.

In 1997 the term 'smartphone' was used for the first time when Ericsson unveiled the concept phone GS88,[6] [7] the first device labelled as 'smartphone'.[8]

Symbian

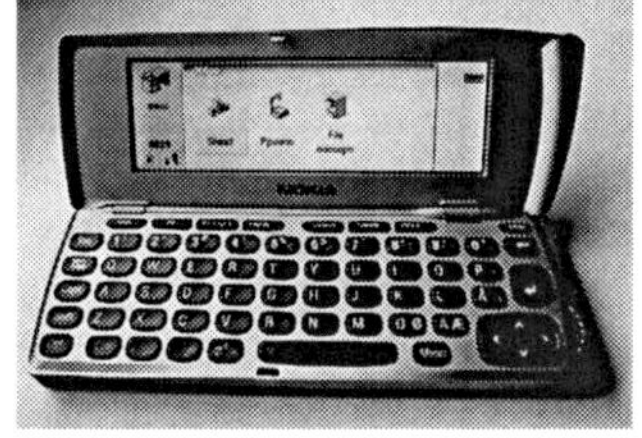

The Nokia 9210 Communicator (Symbian 2000 model smartphone)

In 2000, the touchscreen Ericsson R380 Smartphone was released.[9] It was the first device to use an open operating system, the Symbian OS.[10] It was the first device marketed as a 'smartphone'.[11] It combined the functions of a mobile phone and a personal digital assistant (PDA).[12] In December 1999 the magazine Popular Science appointed the Ericsson R380 Smartphone to one of the most important advances in science and technology.[13] It was a groundbreaking device since it was as small and light as a normal mobile phone.[14] In 2002 it was followed up by P800.[15]

Also in 2000, the Nokia 9210 communicator was introduced, which was the first color screen model from the Nokia Communicator line. It was a true smartphone with an open operating system, the Symbian OS. It was followed by the 9500 Communicator, which also was Nokia's first cameraphone and first Wi-Fi phone. The 9300 Communicator was smaller, and the latest E90 Communicator includes GPS. The Nokia Communicator model is remarkable for also having been the most costly phone model sold by a major brand for almost the full life of the model series, costing easily 20% and sometimes 40% more than the next most expensive smartphone by any major producer.

In 2007 Nokia launched the Nokia N95 which integrated a wide range of multimedia features into a consumer-oriented smartphone: GPS, a 5 megapixel camera with autofocus and LED flash, 3G and Wi-Fi connectivity and TV-out. In the next few years these features would become standard on high-end smartphones. The Nokia 6110 Navigator is a Symbian based dedicated GPS phone introduced in June 2007.

In 2010 Nokia released the Nokia N8 smartphone with a stylus-free capacitive touchscreen, the first device to use the new Symbian^3 OS.[16] It featured a 12 megapixel camera with Xenon flash able to record HD video in 720p, described by Mobile Burn as the best camera in a phone,[17] and satellite navigation that Mobile Choice described as the best on any phone.[18] It also featured a front-facing VGA camera for videoconferencing.

Symbian was the number one smartphone platform by market share from 1996 until 2011 when it dropped to second place behind Google's Android OS. In February 2011, Nokia announced that it would replace Symbian with Windows Phone as the operating system on all of its future smartphones.[19] This transition was completed in October 2011, when Nokia announced its first line of Windows Phone 7.5 smartphones, Lumia 710 and 800.[20]

Palm, Windows, and BlackBerry

In the late 1990s the vast majority of mobile phones had only basic phone features and many people who needed functionality beyond that also carried PDA and/or pager type devices running early versions of operating systems such as Palm OS, BlackBerry OS or Windows CE/Pocket PC.[1] Later versions of these systems started integrating cell phone capabilities with their PDA and messaging features and support of third-party applications. Today, high-end devices running these systems are often branded smartphones.

The HTC Touch Pro2 smartphone (May 2009)

In early 2001, Palm, Inc. introduced the Kyocera 6035, the first smartphone to be deployed in widespread use in the United States. This device combined the features of a personal digital assistant (PDA) with a wireless phone that operated on the Verizon Wireless network. For example, a user could select a name from the PDA contact list, and the device would dial that contact's phone number. The device also supported limited web browsing.[21] The device received a very positive reception from technology publications, but the product line never became widespread outside North America.[22]

In 2001 Microsoft announced its Windows CE Pocket PC OS would be offered as "Microsoft Windows Powered Smartphone 2002."[23] Microsoft originally defined its Windows Smartphone products as lacking a touchscreen and offering a lower screen resolution compared to its sibling Pocket PC devices.

In early 2002 Handspring released the Palm OS Treo smartphone, utilizing a full keyboard that combined wireless web browsing, email, calendar, and contact organizer with mobile third-party applications that could be downloaded or synced with a computer.[24]

In 2002 RIM released their first BlackBerry devices with integrated phone functionality and shifted the positioning of their products from 2-way pagers to email-capable mobile phones. The BlackBerry line evolved into the first smartphone optimized for wireless email use and had achieved a total customer base of about 32 million subscribers by December 2009.[25]

iPhone

The original iPhone (June 2007)

In 2007, Apple Inc. introduced its first iPhone. It was initially costly, priced at $499 for the cheaper of two models on top of a two year contract. The first mobile phone to use a multi-touch interface, the iPhone was notable for its use of a large touchscreen for direct finger input as its main means of interaction, instead of having a stylus, keyboard, and/or keypad, which were the typical input methods for other smartphones at the time. The iPhone featured a web browser that *Ars Technica* then described as "far superior" to anything offered by that of its competitors.[26] Initially lacking the capability to install native applications beyond the ones built-in to its OS, at WWDC in June 2007 Apple announced that the iPhone would support third-party "web 2.0 applications" running in its web browser that share the look and feel of the iPhone interface.[27] As a result of the iPhone's initial inability to install third-party native applications, some reviewers did not consider the originally released device to accurately fit the definition of a smartphone "by conventional terms."[28] A process called jailbreaking emerged quickly to provide unofficial third-party native applications. The different functions of the iPhone (including a GPS unit, kitchen timer, radio, map book, calendar, notepad, and many others) allowed consumers to replace all of these items.[29]

In July 2008, Apple introduced its second generation iPhone with a lower list price starting at $199 and 3G support. Released with it, Apple also created the App Store, adding the capability for any iPhone or iPod Touch to officially execute additional native applications (both free and paid) installed directly over a Wi-Fi or cellular network, without the more typical process at the time of requiring a PC for installation. Applications could additionally be browsed through and downloaded directly via the iTunes software client on Macintosh and Windows PCs, rather than by searching through multiple sites across the Internet. Featuring over 500 applications at launch,[30] Apple's App Store was immediately very popular,[31] quickly growing to become a huge success.[32] [33]

In June 2010, Apple introduced iOS 4, which included APIs to allow third-party applications to multitask,[34] and the iPhone 4, which included a 960×640 pixel display with a pixel density of 326 pixels per inch (ppi), a 5 megapixel camera with LED flash capable of recording HD video in 720p at 30 frames per second, a front-facing VGA camera for videoconferencing, a 1 GHz processor, and other improvements.[35] In early 2011 the iPhone 4 became available through Verizon Wireless, ending AT&T's exclusivity of the handset in the U.S.,[36] [37] [38] and allowing the handset's 3G connection to be used as a wireless Wi-Fi hotspot for the first time, to up to 5 other devices.[39] Software updates subsequently added this capability to other iPhones running iOS 4.[40] [41]

The iPhone 4S was announced on October 4, 2011, improving upon the iPhone 4 with a dual core A5 processor, an 8 megapixel camera capable of recording 1080p video at 30 frames per second, World phone capability allowing it to work on both GSM & CDMA networks, and the Siri automated voice assistant.[42] On October 10, Apple announced that over one million iPhone 4Ss had been pre-ordered within the first 24 hours of it being on sale, beating the 600,000 device record set by the iPhone 4,[43] [44] despite the iPhone 4S failing to impress some critics at the announcement[45] [46] due to their expectations of an "iPhone 5" with rumored drastic changes compared to the iPhone 4 such as a new case design and larger screen.[47] Along with the iPhone 4S Apple also released iOS 5 and iCloud, untethering iOS devices from Macintosh or Windows PCs for device activation, backup, and synchronization,[48] along with additional new and improved features.[49]

There are about 35 percent of Americans that have some sort of smartphone. This shows that the market is spreading fast and there are also more capabilities for smartphones because of this spread.[50]

Smartphones are also mainly valuable based on the operating system. For example, the iPhone runs on the iOS and other devices run different operating systems which makes the functionality of these systems different.[51]

Android

The Android operating system for smartphones was released in 2008. Android is an open-source platform backed by Google, along with major hardware and software developers (such as Intel, HTC, ARM, Motorola and Samsung, to name a few), that form the Open Handset Alliance.[52] The first phone to use Android was the HTC Dream, branded for distribution by T-Mobile as the G1. The software suite included on the phone consists of integration with Google's proprietary applications, such as Maps, Calendar, and Gmail, and a full HTML web browser. Android supports the execution of native applications and a preemptive multitasking capability (in the form of services). Third-party apps are available via the Android Market (released October 2008), including both free and paid apps.

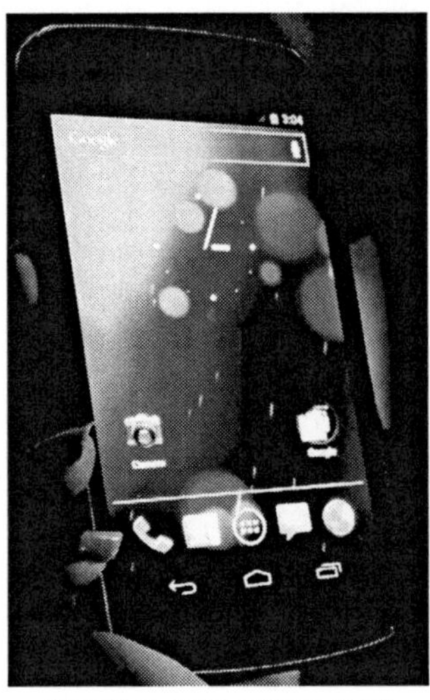

Galaxy Nexus, the latest "Google phone"

In January 2010, Google launched the Nexus One smartphone using its Android OS. Although Android has multi-touch abilities, Google initially removed that feature from the Nexus One,[53] but it was added through a firmware update on February 2, 2010.[54]

Concerning the Xperia Play smartphone, an analyst at CCS Insight said in March 2011 that "Console wars are moving to the mobile platform".[55] In the same month, the HTC EVO 3D was announced by HTC Corporation, which can produce 3D effects with no need for special glasses (autostereoscopy).[56] The HTC EVO 3D was officially released on June 24, 2011.[57]

Bada

The Bada operating system for smartphones was announced by Samsung on 10 November 2009.[58] [59] The first Bada-based phone was the Samsung Wave S8500, released on June 1, 2010,[60] [61] which sold one million handsets in its first 4 weeks on the market.[62]

Samsung shipped 3.5 million phones running Bada in Q1 of 2011.[63] This rose to 4.5 million phones in Q2 of 2011.[64]

Patent licensing and litigation

Recently the number of lawsuits, trade complaints, and countersuits and complaints based on patents and designs in the markets for smartphones, and devices based on smartphone OSes such as Android, has been increasing significantly.

Timeline[65] [66] [67] [68] [69] (initial suits, countersuits, rulings, licence agreements, and other major events in *italics*):

- 2009, Oct 22: *Nokia sues Apple over 10 patents.*[70] [71]
- 2009, Dec 11: *Apple countersues Nokia over 13 patents.*[72]
- 2009, Dec 29: Nokia files a second lawsuit[73] and a U.S. International Trade Commission (ITC) complaint against Apple over 7 more patents.[74]

- 2010, Jan 15: Apple files an ITC complaint against Nokia over 9 patents.[75] [76]
- 2010, Feb 19: Apple drops 4 patents from their countersuit against Nokia that are in their ITC complaint against Nokia.
- 2010, Feb 24: Apple countersues Nokia in Nokia's second lawsuit, over the 9 patents that are in Apple's ITC complaint.
- 2010, Mar 02: *Apple sues HTC over 10 patents and files an ITC complaint against HTC over 10 other patents.*[77] [78] [79] [80] [81]
- 2010, Apr 26: 5 of the patents in Apple's ITC complaint against Nokia are merged into their ITC complaint against HTC.
- 2010, Apr 27: *HTC signs an agreement with Microsoft to licence Microsoft patents in return for royalties on HTC's Android-based devices*[82] [83] (rumored to be $5 per handset).
- 2010, May 7: Nokia files a third lawsuit against Apple over 5 more patents.[84]
- 2010, May 12: *HTC files an ITC complaint against Apple over 5 patents.*[85]
- 2010, May 28: S3 Graphics files an ITC complaint against Apple over 4 patents used in the iPhone, iPod Touch, iPad, and Apple computers.[86]
- 2010, Jun 28: Apple countersues Nokia in Nokia's third lawsuit, over 7 more patents.
- 2010, Jul 06: *HTC countersues Apple over 3 patents.*
- 2010, Jul 21: Nokia drops 1 patent from their ITC complaint against Apple.
- 2010, Aug 12: *Oracle sues Google over 7 patents relating to the use of Java in Android.*[87]
- 2010, Sep 17: Nokia adds 2 more patents to their third lawsuit against Apple.
- 2010, Sep 27: Apple sues Nokia in the UK and Germany over 9 patents.
- 2010, Sep 30: Nokia countersues Apple in Germany over 4 patents.
- 2010, Oct 01: *Microsoft files an ITC complaint and a lawsuit against Motorola over 9 patents.*[88] [89]
- 2010, Oct 06: *Motorola sues Apple over 18 patents, and files an ITC complaint against Apple over 6 of them.*[90]
- 2010, Oct 08: *Motorola files a request for declaratory judgement that they do not infringe 12 Apple patents, and that those patents be declared invalid.*[91] [92]
- 2010, Oct 12: Nokia adds 3 more patents to their countersuit against Apple in Germany.
- 2010, Oct 25: Nokia sues Apple in another German court over 5 patents.
- 2010, Oct 28: Apple drops 4 patents from their ITC complaint against HTC and/or Nokia.
- 2010, Oct 29: *Apple sues Motorola over 6 patents, and files an ITC complaint against Motorola over 3 of them.*[93] [94]
- 2010, Nov 05: HTC drops 1 patent from their ITC complaint against Apple.
- 2010, Nov 09: *Microsoft alleges Motorola has failed to comply with RAND (reasonable and non-discriminatory) licensing obligations.*
- 2010, Nov 10: *Motorola sues Microsoft over 7 patents in one court and 9 patents in another.*
- 2010, Nov 18: *Apple makes counterclaims against Motorola over 6 patents.*
- 2010, Nov 22: *Motorola files an ITC complaint against Microsoft over 5 patents.*
- 2010, Dec 01: *Apple adds the 12 patents to their suit against Motorola that Motorola previously requested declaratory judgement that they do not infringe.*[95]
- 2010, Dec 03: Nokia countersues Apple in the UK over 4 patents, and files a new suit against Apple in the Netherlands over 2 patents.
- 2010, Dec 03: Apple countersues Nokia in Nokia's second German lawsuit, over 1 patent and 2 utility models.
- 2010, Dec 06: Nokia drops 1 patent from their ITC complaint against Apple.
- 2010, Dec 15 and 22: Nokia and Apple take their first German suit/countersuit to the Federal Patent Court of Germany.
- 2010, Dec 23: Motorola files a third lawsuit against Microsoft over 3 patents.
- 2010, Dec 23: Microsoft countersues Motorola over 7 patents.

- 2011, Jan 06: The third Nokia/Apple lawsuit/countersuit is transferred to the location of the first and second ones.
- 2011, Jan 18: Apple seeks to invalidate one Nokia patent in the UK, which it was not yet being sued over.
- 2011, Jan 18: Motorola drops 1 patent from their lawsuits against Microsoft.
- 2011, Jan 19: Microsoft counterclaims against Motorola, asserting 5 patents.
- 2011, Jan 25: Microsoft counterclaims against Motorola, asserting 2 patents.
- 2011, Feb 14: Motorola adds 2 patents to their lawsuits against Microsoft.
- 2011, Feb 22: Apple drops 1 more patent from their ITC complaint against HTC and Nokia.
- 2011, Mar 21: *Microsoft sues Barnes & Noble over the Android operating system in the Nook ebook reader.*[96]
- 2011, Mar 25: *ITC finds that Apple does not infringe on 5 Nokia patents.*
- 2011, Mar 29: Nokia files an ITC complaint against Apple over 7 more patents, and a fourth lawsuit over 6 of those.[97] [98]
- 2011, Apr 15: *Apple sues Samsung for patent and trademark infringement* (7 utility patents, 3 design patents, 3 registered trade dresses, 6 trademarked icons) with its Galaxy line of mobile products, including the *Galaxy S* smartphone and the *Galaxy Tab* tablet.[99] [100]
- 2011, Apr 22: *Samsung sues Apple in South Korea (5 patents), Japan (2 patents), and Germany (3 patents).*[101]
- 2011, Apr 28: *Samsung countersues Apple over 10 patents.*[102]
- 2011, Apr 29: Apple drops 1 more patent from their ITC complaint against HTC.
- 2011, May 18: *Samsung ordered to provide Apple samples of the announced Galaxy S2, Infuse 4G,* and *Infuse 4G LTE* smartphones, as well as the *Galaxy Tab 8.9* and *Galaxy Tab 10.1* tablets as part of Apple's lawsuit against the company.[103] [104]
- 2011, May 18: *Samsung files a court motion for Apple to provide samples of the unannounced iPhone 5 and iPad 3 prototypes.*[105]
- 2011, Jun 14: *Nokia and Apple settle their litigation with Apple agreeing to pay Nokia an undisclosed one-time payment as well as continuing royalties.*[106] [107]
- 2011, Jun 16: *Apple amends its lawsuit against Samsung,* dropping 2 utility patents and 1 design patent, and adding 3 new utility patents plus 4 trade dress applications, *now covering the Samsung Galaxy Tab 10.1*[108]
- 2011, Jun 22: Apple countersues Samsung in South Korea over an unknown number of patents.
- 2011, Jun 22: *Samsung's motion to be provided samples of Apple's unannounced iPhone 5 and iPad 3 prototypes is denied.*[109]
- 2011, Jun 27: General Dynamics Itronix signs an agreement with Microsoft to licence Microsoft patents in return for royalties on General Dynamics Itronix's Android-based devices.[110] [111]
- 2011, Jun 28: Samsung files an ITC complaint and a lawsuit against Apple over 5 patents.
- 2011, Jun 29: Samsung sues Apple in London, UK over an unknown number of patents, and a Samsung lawsuit against Apple in Italy becomes known (details unknown).
- 2011, Jun 29: Velocity Micro signs an agreement with Microsoft to licence Microsoft patents in return for royalties on Velocity Micro's Android-based devices.[111] [112]
- 2011, Jun 30: Samsung converts its countersuit against Apple into counterclaims against Apple's suit, dropping 2 patents but adding 4 more.
- 2011, Jun 30: *A consortium of companies made up of Apple, EMC Corporation, Ericsson, Microsoft, Research In Motion and Sony win against Google*[113] *in an auction of over 6,000 Nortel mobile-related telecommunications patents for $4.5 billion USD.*[114] [115]
- 2011, Jun 30: Onkyo signs an agreement with Microsoft to licence Microsoft patents in return for royalties on Onkyo's Android-based devices.[111] [116]
- 2011, Jul 01: *Apple files for preliminary injunction against 4 Samsung products: Infuse 4G, Galaxy S 4G, Droid Charge,* and *Galaxy Tab 10.1* based on 3 design patents and 1 utility patent.[117]
- 2011, Jul 01: *ITC rules that Apple infringes on 2 patents held by S3 Graphics, while not infringing on 2 others.*[118]

- 2011, Jul 05: *Apple files an ITC complaint against Samsung over 6 smartphones and 2 tablets* infringing 5 utility patents and 2 design patents.
- 2011, Jul 05: Wistron signs an agreement with Microsoft to licence Microsoft patents in return for royalties on Wistron's Android-based devices.[111] [119] [120]
- 2011, Jul 06: *HTC agrees to purchase S3 Graphics* to secure 235 patents for use in its defense against Apple.[121] [122] [123]
- 2011, Jul 06: *Microsoft seeks $15 licensing fees from Samsung for a range of claimed patent violations on every Android device.*[124]
- 2011, Jul 11: Apple files a second ITC complaint against HTC over 5 more patents, and sues HTC over 4 patents from this second ITC complaint that they weren't already suing HTC over.[125] [126]
- 2011, Jul 11-12: Google acquires 1,029 Patents from IBM for an undisclosed amount.[127] [128]
- 2011, Jul 15: *ITC finds HTC infringes on 2 Apple patents.*[129]
- 2011, Jul 29: HTC sues Apple in London, UK over an unknown number of patents.
- 2011, Aug 02: *Apple sues Samsung in Australia over 10 patents, resulting in Samsung delaying the launch and halting advertising of the Samsung Galaxy Tab 10.1 tablet in Australia* to an indefinite date.[130] [131]
- 2011, Aug 09: *A German court issues a preliminary injunction against the Samsung Galaxy Tab 10.1* in Apple's lawsuit against Samsung *which causes its sale to be banned in most of Europe.*[132] [133]
- 2011, Aug 15: *Google announces its intention to buy Motorola Mobility for $12.5 billion USD.* Eighteen of Motorola's patents could potentially be used for defense or countersuits against Apple and Microsoft, and may influence the smartphone war. These patents may change the balance of power, and force the various players to settle their lawsuits.[134] [135]
- 2011, Aug 16: *The Samsung Galaxy Tab 10.1 sales ban in Europe is lifted outside of Germany.*[136] [137]
- 2011, Aug 17: Google acquires 1,023 more patents from IBM for an undisclosed amount (not revealed until 13 Sep 2011).[138]
- 2011, Aug 23: *Microsoft files a complaint with the ITC requesting a ban on several key Motorola smartphones and devices in the USA* based on infringements of 7 patents.[139] [140]
- 2011, Aug 24: *A court in the Netherlands rules that Samsung will be banned from selling the Galaxy S, Galaxy S II and Galaxy Ace in a number of European countries* due to Apple's patent infringement claims.[141]
- 2011, Sep 02: *Apple granted preliminary injunction against Samsung preventing display of the prototype Samsung Galaxy Tab 7.7 tablet at the IFA trade show in Berlin.*[142]
- 2011, Sep 02: *Apple court filings assert that Andy Rubin got inspiration for Android framework while working at Apple* before working at General Magic and Danger, Inc.[143]
- 2011, Sep 07: *HTC countersues Apple using nine patents from Google.* The move is seen as a possible first step for Google giving direct support in lawsuits involving manufacturers using Android.[144] [145] [146] [147]
- 2011, Sep 08: *Acer*[148] *and ViewSonic*[149] *sign patent license agreements with Microsoft* regarding their use of Android on smartphones and tablets.[150] [151]
- 2011, Sep 09: *Apple's preliminary injunction against sales of the Samsung Galaxy Tab 10.1 in Germany is upheld.*[152]
- 2011, Sep 12: Samsung announces a lawsuit against Apple in France that had been filed in July over 3 patents.[153]
- 2011, Sep 12: Apple countersues Samsung in the UK over an unknown number of patents.[154]
- 2011, Sep 13: Google's August 17 acquisition of 1,023 patents from IBM is revealed by the U.S. Patent and Trademark Office.[138] [155]
- 2011, Sep 17: Samsung countersues Apple in Australia over 7 patents.[156]
- 2011, Sep 28: *Samsung signs an agreement with Microsoft to licence Microsoft patents in return for royalties on Samsung's Android-based devices.*[157] [158] [159]

- 2011, Oct 12: *An Australian court issues a preliminary injunction against the Samsung Galaxy Tab 10.1 in Apple's lawsuit against Samsung which prevents its sale in Australia leading up to the 2011 holiday season.*[160]
- 2011, Oct 13: *Quanta signs an agreement with Microsoft to licence Microsoft patents in return for royalties on Quanta's Android and Chrome-based devices.*[161] [162]
- 2011, Oct 13: *Judge in Apple's U.S. lawsuit against Samsung agrees that Samsung's tablets infringe on Apple's patents, but also that the validity of some of the patents might be questionable.*[163]

Screen

Screens on smartphones vary largely in both display size and display resolution. The most common screen sizes range from 2 inches to over 4 inches (measured diagonally). Some 5 inch screen devices exist that run on mobile OSes and have the ability to make phone calls, such as the discontinued Dell Streak and the current Samsung Galaxy Note. Ergonomics arguments have been made that increasing screen sizes start to negatively impact usability.

Common resolutions for smartphone screens vary from 240×320 to 720×1280, with many flagship Android phones at 480×800 or 540×960, the iPhone 4/4S at 640×960 and Galaxy Nexus and HTC Rezound at 720×1280.

Application stores

The introduction of Apple's App Store for the iPhone and iPod Touch in July 2008 popularized manufacturer-hosted online distribution for third-party applications focused on a single platform. Before this, smartphone application distribution was largely dependent on third-party sources providing applications for multiple platforms, such as GetJar, Handango, Handmark, PocketGear, and others.

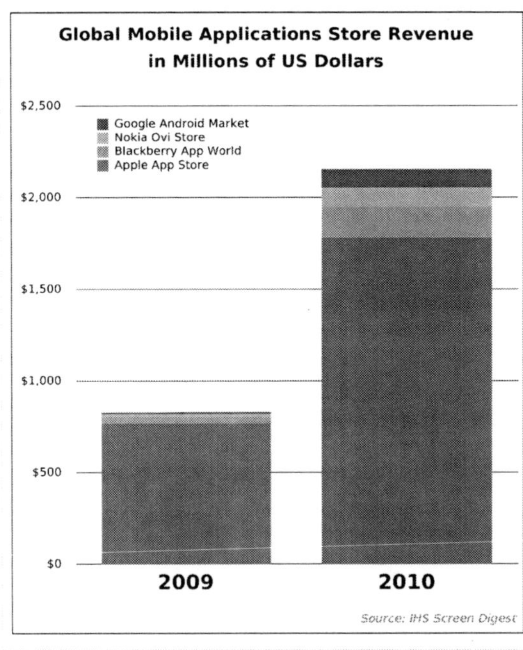

The iPhone's platform is officially restricted to installing apps through the App Store, through "B2B" deployment, and on an "Ad Hoc" basis on up to 100 iPhones.[164] Through jailbreaking it can install apps from other sources. Other platforms may allow application distribution through additional sources outside of their manufacturer-provided app stores, such as third-party app stores and downloads from individual websites.

Following the success of Apple's App Store other smartphone manufacturers quickly launched application stores of their own. Google launched the Android Market in October 2008. RIM launched its app store, BlackBerry App World, in April 2009. Nokia launched its Ovi Store in May 2009. Palm launched its Palm App Catalog for webOS in June 2009. Microsoft launched an application store for Windows Mobile called Windows Marketplace for Mobile in October 2009, and then a separate Windows

Phone Marketplace for Windows Phone in October 2010. Samsung launched Samsung Apps for its Bada based phones in June 2010. Amazon launched its Amazon Appstore for the Google Android operating system in March 2011.

Store	2009 (millions U.S.)	2010 (millions U.S.)[165]
Apple App Store	$769	$1782
Blackberry App World	$36	$165
Nokia Ovi Store	$13	$105
Google Android Market	$11	$102
Total	$828	$2155

The relatively high revenue of U.S. $1782 million in 2010 for Apple's App Store compared to competitor's stores[165] can be attributed to a combination of factors. In large part this can be attributed to having the largest number of apps available and the highest download volume of any mobile app store in 2010, but besides that only 28% of the apps in Apple's App Store were free apps, compared to over 57% in the Android Market. Similarly, Nokia's Ovi Store and the BlackBerry App World both had only 26% of their apps available for free, but both generated higher revenues than the Android Market despite having much lower download volumes.[166]

Malicious software attacks

As smartphone adoption goes up they have increasingly become subject to attacks by malicious software (malware).[167] [168]

Frequently this malware is distributed through application stores that have minimal or no review process for their content.[169] In some cases malware has been hidden in pirated versions of legitimate apps, which are then distributed through 3rd party app stores.[170] [171] Malware risk also comes from what's known as an "update attack," where a legitimate application is later changed to include a malware component, which users then install when they are notified that the app has been updated. Additionally, the ability to acquire software directly from links on the web results in a distribution vector called "malvertizing," where users are directed to click on links, such as on ads that look legitimate, which then open in the device's web browser and cause malware to be downloaded and installed automatically.[172]

Typical smartphone malware leverages platform vulnerabilities that allow it to gain root access on the device in the background. Using this access the malware installs additional software to target communications, location, or other personal identifying information. A common form of malware on mobile phones is the SMS trojan, which sends premium SMS messages, possibly while unknowingly running in the background of a legitimate application. These premium SMS messages run up charges on the owners phone bill which cannot be recovered.

In August 2010, Kaspersky Lab reported detection of the first malicious program for smartphones running on Google's Android operating system, named Trojan-SMS.AndroidOS.FakePlayer.a, an SMS trojan which had already infected a number of devices using that OS.[173] Over the spring of 2011 Android malware increased 76%, according to McAfee.[167] [174] A report from Juniper Global Threat Center notes that malware on the Android platform increased 400% from 2009 to the summer of 2010, and then saw a 472% increase between July and November 2011.[169] The Juniper report indicates that 55% of Android malware acts as spyware, and 44% are SMS trojans.

While there have been and continue to be potential security flaws in iOS,[175] as of at least August 2011 there were no known malware or spyware apps in Apple's App Store, according to security firm Lookout. There are however commercial spyware applications available, outside the App Store, for jailbroken iOS devices.[172] In June 2011 Symantec's 23-page report "A Window Into Mobile Device Security" characterized (non-jailbroken) devices running iOS as having "full protection" against malware attacks.[176]

Symbian and older versions Windows Mobile have had to contend with a degree of malware in the past, but as legacy systems it is believed that the people who previously targeted them have shifted their focus to Android.[169] There were also a few Palm OS viruses.

The only mobile platform other than Apple's iOS without reports of malware so far is HP's (formerly Palm's) webOS, but this may be explained by its relatively low adoption rate.[174]

The best way to reduce a device's vulnerability to malware attacks is to install the most recent versions of operating systems which include security patches. This can be complicated by long delays[177] in software updates for many devices which have had their software modified with custom "skins," services, or promotional on-deck apps by their manufacturer or mobile carrier.[172] In some cases a device may no longer be receiving updates from its manufacturer or carrier, leaving it vulnerable to exploits that have been patched in an OS version that's more recent than the device's last supported one.

Market share

Smartphone market share

For several years, demand for advanced mobile devices boasting powerful processors and graphics processing units, abundant storage (flash memory) for applications and media files, high-resolution screens with multi-touch capability, and open operating systems has outpaced the rest of the mobile phone market.[178]

According to an early 2010 study by ComScore, over 45.5 million people in the United States owned smartphones out of 234 million total subscribers.[179] Despite the large increase in smartphone sales in the last few years, smartphone shipments only made up 20% of total handset shipments as of the first half of 2010.[180]

According to Gartner in their report dated November 2010, total smartphone sales doubled in one year and now smartphones represent 19.3 percent of total mobile phone sales.[181] Smartphone sales increased in 2010 by 72.1 percent from the prior year, whereas sales for all mobile phones only increased by 32%.[182] [183]

According to an Olswang report in early 2011, the rate of smartphone adoption is accelerating: as of March 2011 22% of UK consumers had a smartphone, with this percentage rising to 31% amongst 24- to 35-year-olds.[184]

In March 2011, Berg Insight reported data that showed global smartphone shipments increased 74% from 2009 to 2010.[185]

A survey of mobile users in the United States by Nielsen in Q3, 2011 reports that smartphone ownership has reached 43% of all U.S. mobile subscribers, with the vast majority of users under the age of 44 owning one. In the 25-34 age range smartphone ownership is reported to be at 62%.[186] NPD Group reports that the share of handset sales that were smartphones in Q3, 2011 reached 59% for consumers 18 and over in the U.S.[187]

In profit share worldwide smartphones now far exceed the share of non-smartphones. According to a November 2011 research note from Canaccord Genuity, Apple Inc. holds 52% of the total mobile industry's operating profits, while only holding 4.2% of the global handset market. HTC and RIM similarly only make smartphones and their wordwide profit shares are at 9% and 7%, respectively. Samsung, in second place after Apple at 29%, makes both smartphones and feature phones and doesn't report a breakdown separating their profits between the two kinds of devices, but it can be intuited that a significant portion of that profit comes from their flagship smartphone devices.[188]

Up to the end of November 2011, camera-equipped smartphones took 27 percent of photos, a significant increase from 17 percent last year. Due to the fact that we carry smartphones with us all the time, smartphones have replaced some functions of Point-and-shoot cameras, except the cameras with big optical zoom such as 10x.[189]

Operating system market shares

2010 saw the rapid rise of the Google Android operating system from 4 percent of new deployments in 2009 to 33 percent at the beginning of 2011 making it share the top position with the since long dominating Symbian OS. The smaller rivals include US popular Blackberry OS, iOS, Samsung's recently introduced Bada, HP's heir of Palm webOS and the Microsoft Windows Phone OS which is now supported by Nokia.

Quantity market shares by Gartner (in one year) (new sales)	
Smartphone OS	**Percent**
Android 2009	3.9%
Android 2010	22.7%
Symbian 2009	46.9%
Symbian 2010	37.6%
RIM 2009	19.9%
RIM 2010	16.0%
iOS 2009	14.4%
iOS 2010	15.7%
Windows 2009	8.7%
Windows 2010	4.2%
Other 2009	6.1%
Other 2010	3.8%

Over late 2009 and 2010 Android's smartphone operating system market share increased very rapidly.[181] In the fourth quarter of 2010, Android surpassed Symbian as the most common operating system in smartphones, with 32.9 million units sold versus 31.0 million. Android-equipped phones sold seven times more than in the prior year.[190] According to Canalys, Google's Android operating system, which is offered to phone makers for free, has raced to the top past operating systems by Nokia, Apple, RIM, and Microsoft. In Q1 2011 Google's Android market share was 35 percent, increasing significantly from 10 percent the previous year, while Nokia's Symbian dropped to 26 percent from 46 percent over the same time period.[191] In the UK, which currently has one of the highest penetrations of smartphones in the World, Android achieved 50% market share in October 2011.[192]

Historical sales figures

Figures in millions.

Year	Android (Google)	Blackberry (RIM)	IPhone (Apple)	Linux	Palm/WebOS (Palm/HP)	Symbian (Nokia)	Windows Mobile/Phone (Microsoft)
2007[193]		11.77	3.3	11.76	1.76	77.68	14.7
2008[194]		23.15	11.42	11.26	2.51	72.93	16.5
2009[195]	6.8	34.35	24.89	8.13	1.19	80.88	15.03
2010[196]	67.22	47.45	46.6			111.58	12.38

Enterprise share by operating system

In a worldwide study of 2,300 workers at 1,100 businesses by iPass it was reported that Apple's iPhones have displaced RIM's BlackBerry devices in enterprise adoption in 2011.[197] The share for iPhones increased to 45% from 31.1% in 2010, while the Blackberry share dropped to 32.2% from 34.5% in the previous year. Android phones also increased in share, to 21.3% from 11.3% in 2010, exceeding Symbian for the first time, which dropped to 7.4% from 12.4%. Windows Mobile and all other smartphone OSes also dropped in 2011 compared to 2010.[198]

Customer loyalty by operating system

According to a survey of more than 6,000 smartphone users through 2010 by mobile analytics firm Zokem, the top five loyalty scores for smartphone platforms are the iPhone at 73%, followed by Google's Android at 40%, Samsung's Bada at 33%, RIM's BlackBerry at 30%, and Symbian S60 at 23%. Windows Mobile and Palm follow at 10% each. Customer loyalty gauges the likelihood that the user of a smartphone platform whose contract has expired or who has broken or lost their phone will repurchase another one that uses that same platform.[199] [200]

Manufacturer market shares

From the launch of their Communicator model in 1996 until 2011 Nokia was dominant in the smartphone market, though has more recently been joined by other competitors in the market. Based on a report by Strategy Analytics, Samsung overtook Nokia in smartphone shipments with an estimated 27.8 million units shipped in Q3 2011[201] (Samsung does not publicly disclose the numbers of their smartphone shipments and sales).

Quantity market shares by Strategy Analytics (note that Gartner instead shows Nokia ahead on Apple sales based on Symbian sales only[202]
(new sales)

Manufacturer	Percent
Apple Q2 2010	13.5%
Apple Q2 2011	18.5%
Samsung Q2 2010	5.0%
Samsung Q2 2011	17.5%
Nokia Q2 2010	38.1%
Nokia Q2 2011	15.2%
Others Q2 2010	43.4%
Others Q2 2011	48.9%

Market share among smartphone manufacturers does not resemble smartphone OS market share numbers due to the differences between the two major smartphone OS sales models: single manufacturer and licensed. Apple's iPhone, Nokia's Symbian, and RIM's BlackBerry smartphones are currently only available from single manufacturers.

Google's Android OS and Microsoft's mobile OSes are platforms that are licensed and used by a variety of manufacturers. As a result, manufacturers of smartphones using licensed OSes all split the total market share of that OS between them, while the total share for a single-manufacturer OS is held by that manufacturer alone.

Note that Nokia's Symbian OS was previously available from several manufacturers under a licensed model, then later predominantly only by Nokia itself more like a single manufacturer model.

Samsung smartphones use a diverse portfolio of operating systems, including their own Bada operating system along with Android and Windows Mobile.[203]

Apple surpassed Nokia worldwide by revenue and profit for the first time in Q2 2011 (though not in market share), with Apple's profit share of the total worldwide smartphone market increasing to 66.3% while Nokia reported a loss.[204]

Between Q2 2010 and Q2 2011 Nokia's worldwide Symbian smartphone sales dropped significantly from 38.1 percent to 15.2 percent, while Samsung smartphone sales increased significantly worldwide from 5% to 17.5%.[205] As of Q1 2011, Nokia had already announced plans to switch to Windows Phone.

Smartphone Customer Satisfaction by J.D. Power and Associates

Manufacturer	Score
Apple 2010	810
Apple 2011	838
HTC 2010	727
HTC 2011	801
Industry Average 2010	753
Industry Average 2011	788
Samsung 2010	724
Samsung 2011	777
Motorola 2010	N/A
Motorola 2011	775
RIM 2010	741
RIM 2011	762
LG 2010	N/A
LG 2011	760
HP/Palm 2010	712
HP/Palm 2011	733
Nokia 2010	720
Nokia 2011	721

Rankings are based on a possible top score of 1000

Nokia still remains the number one company in the worldwide mobile phone market with sales for Q2 2011 of 88.5 million when including feature phone platforms such as S40, compared with 16.7 million smartphones running Symbian.[206]

According to Nielsen in July 2011, in the United States Apple is the top smartphone manufacturer at 28% of the market, with RIM at 20%. Google Android has 39% of the U.S. market as a whole, but this is split between HTC at 14%, Motorola at 11%, Samsung at 8%, and other remaining manufacturers at 6%. HTC's total share of the U.S.

smartphone market actually ties RIM at 20%, since sales of their smartphones running Microsoft's mobile operating systems account for 6% of the total market. Samsung similarly gains 2% of overall U.S. market share due to their sales of Microsoft OS-based smartphones. In contrast to the worldwide market, Nokia's share of U.S. smartphone sales is very small, at only 2%.[207] [208] Nielsen's Q3, 2011 survey of mobile users maintains Apple as the top U.S. smartphone maker with a continued 28% of the market, with RIM dropping from 20% to 18%.[186] While Google Android increased in total operating system share from 39% to 43% of the U.S. market, it remains fragmented amongst many different manufacturers. Over the same quarter Microsoft managed a modest gain from 6% to 7% total U.S. smartphone OS share.

Checks with U.S. carriers by technology analyst firm Canaccord Genuity in April and August 2011 have found that Apple's iPhone 4 has consistently been the top selling device at AT&T and Verizon. In addition, the second most popular spot at AT&T has been maintained by the iPhone 3GS, which was originally released in 2009 (and has never been sold on Verizon). In August 2011 the most popular smartphones on Sprint and T-Mobile in the U.S. were the HTC EVO 3D 4G and HTC Sensation, respectively. The other second most popular smartphones were the Samsung Charge 4G on Verizon, the Motorola Photon 4G on Sprint, and the HTC myTouch 4G Slide on T-Mobile.[209] [210] NPD Group reported that in Q3, 2011 the overall top 5 smartphones by sales across all carriers in the U.S. were, in order: the iPhone 4, iPhone 3GS, HTC EVO 4G, Motorola Droid 3, and Samsung Intensity II.[187]

Currently the vast majority of smartphones are manufactured in China, Taiwan and Mexico, for companies based in the U.S. (Apple, HP, Motorola), South Korea (LG, Samsung), Canada (RIM), Finland (Nokia), Taiwan (HTC) and the U.K. (Sony Ericsson).

In Q3 2011, Samsung became the world's number one smartphone vendor, selling 24 million smartphones. Nokia's Symbian platform remained in second place with 19.5 million Symbian smartphones, and Apple fell to third place, with 17 million phones. Android had 52.5% of the market, with 60.5 million Android phones being sold. In the US, HTC became the largest smartphone vendor.[211]

Customer satisfaction by manufacturer

According to global marketing information services firm J.D. Power and Associates smartphones from Apple Inc. have been consistently[212] ranking highest in customer satisfaction,[213] [214] with a late 2011 score of 838 out of 1000. Based on the responses to their most recent survey of 6,898 smartphone users, Apple was followed in ranking by HTC (801), Samsung (777), Motorola (775), RIM (762), LG (760), Palm (733), and Nokia (721).[215] [216] [217] [218]

Open-source development

The open-source culture has penetrated the smartphone market in several ways. There have been attempts to create open source hardware and software for smartphones.

In February 2010 Nokia made Symbian open source. Thus, most commercial smartphones were based on open-source operating systems. These include those based on Linux, such as Google's Android, Nokia's Maemo, Hewlett-Packard's webOS, and those based on BSD, such as the Darwin-based Apple iOS. Maemo was later merged with Intel's project Moblin to form MeeGo.[219] [220]

Popular services

Location-based check-in services

According to a ComScore report released on May 12, 2011, nearly one in five smartphone users are tapping into check-in services like Foursquare and Gowalla. A total of 16.7 million mobile-phone subscribers used location-based services on their phones in March 2011.[221]

Second screen

The smartphones have introduced a new way of watching television.[222] The second screen is a consequence of the media multitasking which is exploding.[223]

See also

- Camera phone and videophone
- Comparison of smartphones
- List of digital distribution platforms for mobile devices
- Mobile broadband connectivity
- Mobile Internet device (MID) and personal digital assistant (PDA)
- Mobile operating system
- Second screen

References

[1] "Smartphone" (http://www.phonescoop.com/glossary/term.php?gid=131). *Phone Scoop*. . Retrieved 2011-12-15.
[2] "Feature Phone" (http://www.phonescoop.com/glossary/term.php?gid=310). *Phone Scoop*. . Retrieved 2011-12-15.
[3] Andrew Nusca (20 August 2009). "Smartphone vs. feature phone arms race heats up; which did you buy?" (http://www.zdnet.com/blog/gadgetreviews/smartphone-vs-feature-phone-arms-race-heats-up-which-did-you-buy/6836). ZDNet. . Retrieved 2011-12-15.
[4] "Smartphone definition from PC Magazine Encyclopedia" (http://www.pcmag.com/encyclopedia_term/0,2542,t=Smartphone&i=51537,00.asp). *PC Magazine*. . Retrieved 2011-12-15.
[5] Schneidawind, J: "Big Blue unveiling", *USA Today*, November 23, 1992, p. 2B.
[6] "Ericsson GS88 Preview" (http://pws.prserv.net/Eri_no_moto/GS88_Preview.htm). *Eri-no-moto*. . Retrieved 2011-12-15.
[7] "History" (http://www.stockholmsmartphone.org/history/). Stockholm Smartphone. . Retrieved 2011-12-15.
[8] "Ericsson GS88 box" (http://www.stockholmsmartphone.org/wp-content/uploads/penelope-box.jpg). . Retrieved 2011-12-15.
[9] "PDA Review: Ericsson R380 Smartphone" (http://www.geek.com/hwswrev/pda/ericr380/). *Geek.com*. . Retrieved 2011-12-15.
[10] "Symbian Device – The OS Evolution" (http://www.i-symbian.com/wp-content/uploads/2009/11/Symbian_Evolution.pdf) (PDF). *Independent Symbian Blog*. . Retrieved 2011-12-15.
[11] "Ericsson Introduces The New R380e" (http://www.mobilemag.com/2001/09/25/ericsson-introduces-the-new-r380e). *Mobile Magazine*. . Retrieved 2011-12-15.
[12] Brown, Bruce. "Ericsson R380 World, Review & Rating" (http://www.pcmag.com/article2/0,2817,40827,00.asp). *PCMag.com*. . Retrieved 2011-12-15.
[13] *Popular Science, December 1999* (http://books.google.com/books?id=8qSgh_Q-YOkC). *Google Books*. . Retrieved 2011-12-15.
[14] "Ericsson R380 PDA & Phone" (http://www.cellular.co.za/ericsson_r380.htm). *CellularOnline*. . Retrieved 2011-12-15.
[15] "Sony Ericsson P800" (http://www.allaboutsymbian.com/features/item/Sony_Ericsson_P800.php). *All About Symbian*. . Retrieved 2011-12-15.
[16] "Nokia N8 smartphone official site" (http://www.nokia.co.uk/gb-en/products/phone/n8-00/). . Retrieved 2011-12-15.
[17] "Nokia N8 review - best camera phone ever" (http://www.mobileburn.com/review.jsp?Id=11093). Mobileburn.com. 2010-10-05. . Retrieved 2011-12-15.
[18] "Mobile Choice Award: Best Sat-Nav" (http://www.mobilechoiceuk.com/News/mobile+choice+award:+Best+Sat-Nav/5274). Mobile Choice. 22 October 2010. . Retrieved 2011-12-15.
[19] "Nokia, Microsoft in pact to rival Apple, Google - Technology & Science" (http://www.cbc.ca/news/technology/story/2011/02/11/nokia-microsoft-smart-phone-apple-google.html). Associated Press. CBC.ca. 2011-02-11. . Retrieved 2011-12-15.
[20] Velazco, Chris (26 October 2011). "Nokia Debuts Their First Windows Phones: The Lumia 800 and Lumia 710" (http://techcrunch.com/2011/10/26/nokia-debuts-lumia-710-and-lumia-800/). *TechCruch*. . Retrieved 26 October 2011.

[21] "Kyocera QCP 6035 Smartphone Review" (http://www.palminfocenter.com/view_story.asp?ID=1707). Palminfocenter.com. 2001-03-16. . Retrieved 2011-09-07.
[22] Segan, Sascha (2010-03-23). "Kyocera Launches First Smartphone In Years | News & Opinion" (http://www.pcmag.com/article2/0,2817,2361664,00.asp). PCmag.com. . Retrieved 2011-09-07.
[23] "Better Living through Software: Microsoft Advances for the Home Highlighted at Consumer Electronics Show 2002: Microsoft showcases new and improved products -- including "Freestyle," "Mira" and Ultimate TV at International CES 2002" (http://www.microsoft.com/presspass/features/2002/Jan02/01-08msces.mspx). Microsoft.com. 2002-01-08. . Retrieved 2011-09-07.
[24] Stephen H. Wildstrom (November 30, 2001). "Handspring's Breakthrough Hybrid" (http://www.businessweek.com/bwdaily/dnflash/nov2001/nf20011129_0157.htm). Businessweek.com. . Retrieved 2011-12-15.
[25] Kevin McLaughlin (December 17, 2009). "BlackBerry Users Call For RIM To Rethink Service" (http://www.crn.com/news/client-devices/222002587/blackberry-users-call-for-rim-to-rethink-service.htm). CRN.com. . Retrieved 2011-12-15.
[26] "iPhone in depth: the Ars review" (http://arstechnica.com/apple/reviews/2007/07/iphone-review.ars/6). *ArsTechnica*. Condé Nast. 9 July 2007. p. 6. . Retrieved 3 August 2010.
[27] "iPhone to Support Third-Party Web 2.0 Applications" (http://www.apple.com/pr/library/2007/06/11iphone.html). *Press Release*. Apple Inc.. June 11, 2007. . Retrieved December 15, 2008.
[28] "The iPhone is not a smartphone" (http://www.engadget.com/2007/01/09/the-iphone-is-not-a-smartphone/). Engadget.com. 9 January 2007. . Retrieved 11 July 2010.
[29] "Smartphones Can Replace These Everyday Items" (http://simpleorganizedlife.com/smartphones-can-replace-these-everyday-items/). Simple. Organized. Life.. October 10, 2011. . Retrieved 2011-12-15.
[30] "iPhone 3G on Sale Tomorrow" (http://www.apple.com/pr/library/2008/07/10iphone.html). *Press Release*. Apple Inc.. 2008-07-10. . Retrieved 2009-01-17.
[31] "iPhone App Store Downloads Top 10 Million in First Weekend" (http://www.apple.com/pr/library/2008/07/14iPhone-App-Store-Downloads-Top-10-Million-in-First-Weekend.html). *Press Release*. Apple Inc.. July 14, 2008. . Retrieved 2011-12-15.
[32] "Apple's App Store Downloads Top 1.5 Billion in First Year" (http://www.apple.com/pr/library/2009/07/14Apples-App-Store-Downloads-Top-1-5-Billion-in-First-Year.html). *Press Release*. Apple Inc.. July 14, 2009. . Retrieved 2011-12-15.
[33] "Apple's App Store Downloads Top 15 Billion" (http://www.apple.com/pr/library/2011/07/07Apples-App-Store-Downloads-Top-15-Billion.html). *Press Release*. Apple Inc.. July 7, 2011. . Retrieved 2011-12-15.
[34] "Apple now accepting iOS 4 apps, multitasking ahoy" (http://www.engadget.com/2010/06/11/apple-now-accepting-ios-4-apps-multitasking-ahoy/). Engadget.com. 11 June 2010. . Retrieved 2011-12-15.
[35] "Apple Presents iPhone 4" (http://www.apple.com/pr/library/2010/06/07Apple-Presents-iPhone-4.html). *Press Release*. Apple Inc.. 2010-06-07. . Retrieved 2011-07-05.
[36] "Liveblog: The Verizon iPhone" (http://voices.washingtonpost.com/fasterforward/2011/01/liveblog_the_verizon_iphone.html). *The Washington Post*. .
[37] Memmott, Mark (2011-01-11). "It's Official: Verizon Has The iPhone 4 : The Two-Way" (http://www.npr.org/blogs/thetwo-way/2011/01/11/132833078/its-official-verizon-has-iphone-4). NPR. . Retrieved 2011-09-07.
[38] Raice, Shayndi (January 12, 2011). "Verizon Unwraps iPhone" (http://online.wsj.com/article/SB10001424052748703791904576075681886276172.html?mod=googlenews_wsj). *The Wall Street Journal*. .
[39] Glenn Fleishman (2011-02-22). "Using the Personal Hotspot on your Verizon iPhone" (http://www.macworld.com/article/158058/2011/02/personal_hotspot_verizon.html). Macworld.com. . Retrieved 2011-03-12.
[40] "iOS 4.3 Software Update" (http://www.apple.com/ios/). Apple Inc.. . Retrieved 2011-03-12.
[41] Dan Moren (2011-03-11). "Hands on with iOS 4.3" (http://www.macworld.com/article/158483/2011/03/firstlook_43.html). Macworld.com. . Retrieved 2011-03-12.
[42] "Apple Launches iPhone 4S, iOS 5 & iCloud" (http://www.apple.com/pr/library/2011/10/04Apple-Launches-iPhone-4S-iOS-5-iCloud.html). Apple Inc.. October 4, 2011. . Retrieved 2011-12-15.
[43] "iPhone 4S Pre-Orders Top One Million in First 24 Hours" (http://www.apple.com/pr/library/2011/10/10iPhone-4S-Pre-Orders-Top-One-Million-in-First-24-Hours.html). Apple. . Retrieved 10 October 2011.
[44] Anderson, Ash. "iPhone 4S Sells 1 Million in Under 24 Hours" (http://www.keynoodle.com/iphone-4s-sells-1-million-in-under-24-hours/). *KeyNoodle*. . Retrieved 2011-12-15.
[45] "Apple's fall from grace" (http://www.apple.com/pr/library/2011/10/04Apple-Launches-iPhone-4S-iOS-5-iCloud.html). BGR.com. October 5, 2011. . Retrieved 2011-12-15.
[46] Michael E. Cohen (October 13, 2011). "iPhone 4S: A Very Palpable Hit" (http://tidbits.com/article/12554). tidbits.com. . Retrieved 2011-12-15.
[47] David Pogue (October 11, 2011). "New iPhone Conceals Sheer Magic" (http://www.nytimes.com/2011/10/12/technology/personaltech/iphone-4s-conceals-sheer-magic-pogue.html?_r=2&pagewanted=all). The New York Times. . Retrieved 2011-12-15.
[48] "Press Info - Apple to Launch iCloud on October 12" (http://www.apple.com/pr/library/2011/10/04Apple-to-Launch-iCloud-on-October-12.html). Apple. 2011-10-04. . Retrieved 2012-01-05.
[49] "Press Info - New Version of iOS Includes Notification Center, iMessage, Newsstand, Twitter Integration Among 200 New Features" (http://www.apple.com/pr/library/2011/06/06New-Version-of-iOS-Includes-Notification-Center-iMessage-Newsstand-Twitter-Integration-Among-200-New-Features.html). Apple.

2011-06-06. . Retrieved 2012-01-05.
[50] Michael B. Farrell (November 12, 2011). "No cash, card? No problem: Paying by smartphone is an emerging trend" (http://articles.boston.com/2011-11-12/business/30391540_1_smartphone-mobile-commerce-card-readers). boston.com. . Retrieved 2011-12-15.
[51] 13 November 2011. (http://cellphones.about.com/od/smartphonebasics/a/what_is+_smart.htm)
[52] "Alliance Members" (http://www.openhandsetalliance.com/oha_members.html). *Open Handset Alliance*. . Retrieved 16 January 2011.
[53] "Weighing Nexus over iPhone – a practical review for the everyday user!" (http://techietrick.blogspot.com/2010/01/weighing-nexus-over-iphone-practical.html). Techietrick.blogspot.com. 2010-01-06. . Retrieved 2011-09-07.
[54] "Nexus One gets a software update, enables multitouch (updated with video!)" (http://www.engadget.com/2010/02/02/nexus-one-gets-a-software-update-enables-multitouch). Engadget.com. . Retrieved 2011-09-07.
[55] Tarmo Virki (2011-02-14). "Sony takes gaming console war to phones" (http://in.reuters.com/article/2011/02/13/idINIndia-54865020110213). Reuters. .
[56] "HTC EVO 3D" (http://web.archive.org/web/20110511105547/http://www.htc.com/www/product/evo3d/overview.html). htc.com. Archived from the original (http://www.htc.com/www/product/evo3d/overview.html) on 2011-05-11. . Retrieved 2011-12-15.
[57] Savov, Vlad. "HTC Evo 3D Launches June 24th" (http://www.engadget.com/2011/06/06/htc-evo-3d-launches-on-june-24th-for-200-joined-by-evo-view-4g/). Engadget.com. . Retrieved 7 June 2011.
[58] Ed Hansberry (11 November 2009). "Samsung Bailing on Windows Mobile" (http://www.informationweek.com/blog/main/archives/2009/11/samsung_bailing.html). *InformationWeek*. .
[59] "Samsung to Discard Windows Phone" (http://www.telecomskorea.com/market-8281.html). *Telecoms Korea*. 9 November 2009. .
[60] "Samsung Wave, first Bada smartphone hits the market" (http://www.bada.com/samsung-wave-first-bada-smartphone-hits-the-market/). *Bada*. 24 May 2010. . Retrieved 3 February 2011.
[61] "BadaWave" (http://badawave.com/). BadaWave. . Retrieved 2012-01-05.
[62] "Samsung Waves away a million" (http://www.theinquirer.net/inquirer/news/1722287/samsung-waves-away-million). The Inquirer. 13 July 2010. .
[63] "Android increases smart phone market leadership with 35% share" (http://www.canalys.com/newsroom/android-increases-smart-phone-market-leadership-35-share). Canalys.com. 2011-05-04. . Retrieved 2011-09-07.
[64] "Samsung Bada shipments up 355% to 4.5 million units in Q2 2011 | asymco news | PG.Biz" (http://www.pocketgamer.co.uk/r/PG.Biz/asymco+news/news.asp?c=32049). Pocket Gamer. . Retrieved 2011-09-07.
[65] Florian Mueller. "Apple vs Android 10.12.02" (http://www.scribd.com/doc/44759893/Apple-vs-Android-10-12-02). . Retrieved 2011-09-08.
[66] Florian Mueller. "NokiaVsApple_11.03.31.100" (http://www.scribd.com/doc/52195210/NokiaVsApple-11-03-31-100). . Retrieved 2011-09-08.
[67] Florian Mueller. "Microsoft vs Motorola 11.04.09" (http://www.scribd.com/doc/52754079/Microsoft-vs-Motorola-11-04-09). . Retrieved 2011-09-08.
[68] Florian Mueller. "AppleVsSamsung_11.07.05" (http://www.scribd.com/doc/59474521/AppleVsSamsung-11-07-05). . Retrieved 2011-09-08.
[69] Florian Mueller. "AppleVsHTCandS3_11.07.29" (http://www.scribd.com/doc/61387056/AppleVsHTCandS3-11-07-29). . Retrieved 2011-09-08.
[70] "Nokia sues Apple over iPhone's use of patented wireless standards" (http://www.appleinsider.com/articles/09/10/22/nokia_sues_apple_over_iphones_use_of_patented_wireless_standards.html). Appleinsider.com. 2009-10-22. . Retrieved 2012-01-05.
[71] "Nokia vs. Apple: the in-depth analysis" (http://www.engadget.com/2009/10/29/nokia-vs-apple-the-in-depth-analysis/). Engadget.com. . Retrieved 2012-01-05.
[72] "Apple countersues Nokia for infringing 13 patents" (http://www.engadget.com/2009/12/11/apple-countersues-nokia-for-infringing-13-patents/). Engadget.com. . Retrieved 2012-01-05.
[73] Foresman, Chris (2010-01-04). "Nokia adds additional lawsuit in patent catfight with Apple" (http://arstechnica.com/apple/news/2010/01/nokia-adds-additional-lawsuit-in-patent-catfight-with-apple.ars). Arstechnica.com. . Retrieved 2012-01-05.
[74] Foresman, Chris (2009-12-29). "Nokia hurls new salvo in spat with Apple, complains to ITC" (http://arstechnica.com/apple/news/2009/12/nokia-hurls-new-salvo-in-spat-with-apple-complains-to-itc.ars). Arstechnica.com. . Retrieved 2012-01-05.
[75] Decker, Susan (2010-01-16). "Apple Files New Trade Complaint With Against Nokia" (http://www.bloomberg.com/apps/news?pid=newsarchive&sid=ao_5HVbD_IRM). Bloomberg.com. . Retrieved 2012-01-05.
[76] Cheng, Jacqui (2010-01-18). "Apple wants Nokia's US imports blocked" (http://arstechnica.com/apple/news/2010/01/apple-continues-nokia-patent-spat-with-its-own-itc-complaint.ars). Arstechnica.com. . Retrieved 2012-01-05.
[77] "Apple vs HTC: a patent breakdown" (http://www.engadget.com/2010/03/02/apple-vs-htc-a-patent-breakdown/). Engadget.com. 2010-03-02. . Retrieved 2012-01-05.
[78] "Google backs HTC in what could be 'long and bloody battle' with Apple" (http://www.appleinsider.com/articles/10/03/03/google_backs_htc_in_what_could_be_long_and_bloody_battle_with_apple.html). Appleinsider.com. 2010-03-03. . Retrieved 2012-01-05.
[79] Bilton, Nick (2010-03-02). "What Apple vs. HTC Could Mean" (http://bits.blogs.nytimes.com/2010/03/02/what-apple-vs-htc-could-mean/). Bits.blogs.nytimes.com. . Retrieved 2012-01-05.
[80] "HTC says it uses own technology, not Apple's" (http://www.macworld.com/article/146819/2010/03/htc_defense.html). Macworld.com. . Retrieved 2012-01-05.

[81] Foresman, Chris (2010-03-09). "HTC lawsuit came after warning by Apple to handset makers" (http://arstechnica.com/apple/news/2010/03/htc-lawsuit-came-after-warning-by-apple-to-handset-makers.ars). Arstechnica.com. . Retrieved 2012-01-05.
[82] "Microsoft Announces Patent Agreement With HTC" (http://www.microsoft.com/Presspass/press/2010/apr10/04-27MSHTCPR.mspx). Microsoft.com. 2010-04-27. . Retrieved 2012-01-05.
[83] "HTC licenses Microsoft's mobile patents for Android phones – First Take" (http://gartenberg.wordpress.com/2010/04/28/htc-licenses-microsofts-mobile-patents-for-android-phones-first-take/). Gartenberg.wordpress.com. 2010-04-28. . Retrieved 2012-01-05.
[84] Johnston, Casey (2010-05-07). "Nokia applies patent thumbscrews to Apple's iPad" (http://arstechnica.com/apple/news/2010/05/nokia-applies-patent-thumbscrews-to-apples-ipad.ars). Arstechnica.com. . Retrieved 2012-01-05.
[85] "HTC countersues Apple, claims infringement of five patents" (http://www.appleinsider.com/articles/10/05/12/htc_countersues_apple_claims_infringement_of_five_patents.html). Appleinsider.com. 2010-05-12. . Retrieved 2012-01-05.
[86] "S3 Graphics Files New 337 Complaint Regarding Certain Electronic Devices With Image Processing Systems" (http://www.itcblog.com/20100602/s3-graphics-files-new-337-complaint-regarding-certain-electronic-devices-with-image-processing-systems/). Itcblog.com. 2010-06-02. . Retrieved 2012-01-05.
[87] August 12, 2010 (2010-08-12). "Oracle sues Google over Android" (http://venturebeat.com/2010/08/12/oracle-sues-google-over-android/). Venturebeat.com. . Retrieved 2012-01-05.
[88] Protalinski, Emil (2010-10-02). "Microsoft sues Motorola, citing Android patent infringement" (http://arstechnica.com/microsoft/news/2010/10/microsoft-sues-motorola-citing-android-patent-infringement.ars). Arstechnica.com. . Retrieved 2012-01-05.
[89] "Microsoft Files Patent Infringement Action Against Motorola" (http://www.microsoft.com/presspass/press/2010/oct10/10-01statement.mspx). Microsoft.com. 2010-10-01. . Retrieved 2012-01-05.
[90] Foresman, Chris (2010-10-06). "Motorola asks ITC, two federal courts to throw book at Apple" (http://arstechnica.com/apple/news/2010/10/motorola-asks-itc-two-federal-courts-to-throw-book-at-apple.ars). Arstechnica.com. . Retrieved 2012-01-05.
[91] "Motorola Moves Offensively with a New Lawsuit against Apple" (http://www.patentlyapple.com/patently-apple/2010/10/motorola-moves-offensively-with-a-new-lawsuit-against-apple.html). Patentlyapple.com. 2010-10-15. . Retrieved 2012-01-05.
[92] Motorola seeks to invalidate Apple phone patents (http://arstechnica.com/apple/news/2010/10/motorola-horns-in-on-apple-vs-htc-files-suit-against-apple.ars).
[93] Apple Files Lawsuit against Motorola to Defend Multi-Touch (http://www.patentlyapple.com/patently-apple/2010/10/apple-files-lawsuit-against-motorola-to-defend-multi-touch.html).
[94] Apple Patent Case Against Motorola to Get Trade Agency Review (http://www.businessweek.com/news/2010-11-23/apple-patent-case-against-motorola-to-get-trade-agency-review.html).
[95] Apple vs. Motorola: now 42 patents-in-suit (24 Apple and 18 Motorola patents) (http://fosspatents.blogspot.com/2010/12/apple-vs-motorola-now-42-patents-in.html).
[96] Microsoft sues Barnes & Noble over Android in Nook (http://www.geekwire.com/2011/breaking-microsoft-sues-barnes-noble-android).
[97] Nokia files second ITC complaint against Apple (http://press.nokia.com/2011/03/29/nokia-files-second-itc-complaint-against-apple/).
[98] FOSS Patents: Escalation: Nokia files new ITC complaint against Apple (plus a corresponding federal lawsuit) (http://fosspatents.blogspot.com/2011/03/escalation-nokia-files-new-itc.html).
[99] "Apple to Samsung: Stop stealing ideas" (http://www.cnn.com/2011/TECH/innovation/04/19/apple.samsung.lawsuit.wired/). *CNN.com*. . Retrieved 19 April 2011.
[100] Apple sues Samsung: a complete lawsuit analysis (http://thisismynext.com/2011/04/19/apple-sues-samsung-analysis/).
[101] Samsung Countersues Apple for Patent Infringement (http://www.pcmag.com/article2/0,2817,2383964,00.asp).
[102] Samsung sues Apple for infringing 10 patents: a closer look (http://thisismynext.com/2011/04/29/samsung-sues-apple-infringing-10-patents-closer/).
[103] Samsung Must Show Cell Phones to Apple (http://www.courthousenews.com/2011/05/19/36708.htm).
[104] Judge Orders Samsung to Give Apple Unreleased Tablets, Smartphones (http://www.pcmag.com/article2/0,2817,2385861,00.asp).
[105] Samsung's lawyers demand to see the iPhone 5 and iPad 3 (http://thisismynext.com/2011/05/28/samsung-apple-iphone-5-ipad-3/).
[106] Nokia Wins Apple Patent-License Deal Cash, Settles Lawsuits (http://www.bloomberg.com/news/2011-06-14/nokia-apple-payments-to-nokia-settle-all-litigation.html).
[107] Apple Settles With Nokia In Patent Lawsuit (http://www.huffingtonpost.com/2011/06/14/apple-nokia-patent-lawsuit-settlement_n_876499.html).
[108] Staff Reporter, "Apple alleges Samsung of 'slavishly copying' its technology; questions new Galaxy Tab 10.1" (http://www.ibtimes.com/articles/165405/20110619/apple-lawsuit-upgrade-samsung-galaxy-tab-10-1-infringement-intellectual-property-rights-ipad-2.htm), *International Business Times*, 19 June 2011.
[109] Shocker: Samsung not allowed to see iPhone 5 and iPad 3 (http://thisismynext.com/2011/06/22/shocker-samsung-allowed-iphone-5-ipad-3/).
[110] "Microsoft and General Dynamics Itronix Sign Patent Agreement: Agreement will cover General Dynamics Itronix devices running the Android platform" (http://www.microsoft.com/Presspass/press/2011/jun11/06-27ItronixPR.mspx). Microsoft.com. 2011-06-27. . Retrieved 2012-01-05.
[111] FOSS Patents: Android device makers sign up as Microsoft patent licensees -- an overview of the situation (http://fosspatents.blogspot.com/2011/06/android-device-makers-sign-up-as.html).

[112] "Microsoft and Velocity Micro, Inc., Sign Patent Agreement Covering Android-Based Devices: Agreement provides broad coverage of Microsoft's patent portfolio" (http://www.microsoft.com/Presspass/press/2011/jun11/06-29VelocityMicroPR.mspx). Microsoft.com. 2011-06-29. . Retrieved 2012-01-05.

[113] Dealtalk: Google bid "pi" for Nortel patents and lost (http://www.reuters.com/article/2011/07/02/us-dealtalk-nortel-google-idUSTRE76104L20110702).

[114] Nortel Announces the Winning Bidder of Its Patent Portfolio for a Purchase Price of US$4.5 Billion (http://www.marketwatch.com/story/nortel-announces-the-winning-bidder-of-its-patent-portfolio-for-a-purchase-price-of-us45-billion-2011-06-30?reflink=MW_news_stmp).

[115] Who Won The 6,000+ Nortel Patents? Apple, RIM, Microsoft — Everyone But Google (http://techcrunch.com/2011/07/01/apple-microsoft-rim-google-nortel-patents/).

[116] "Microsoft and Onkyo Corp. Sign Patent Agreement Covering Android-Based Tablets: Agreement provides broad coverage of Microsoft's patent portfolio" (http://www.microsoft.com/Presspass/press/2011/jun11/06-30OnkyoPR.mspx?rss_fdn=Custom). Microsoft.com. . Retrieved 2012-01-05.

[117] Apple files motion for preliminary injunction in the U.S. against four Samsung products: Infuse 4G, Galaxy S 4G, Droid Charge, Galaxy Tab 10.1 (http://fosspatents.blogspot.com/2011/07/apple-files-motion-for-preliminary.html).

[118] ITC ruling mixed in S3 Graphics v. Apple (http://news.cnet.com/8301-27076_3-20076219-248/itc-ruling-mixed-in-s3-graphics-v-apple/?tag=mncol;txt).

[119] "Microsoft and Wistron Sign Patent Agreement: Agreement will cover Wistron's Android tablets, smartphones and e-readers" (http://www.microsoft.com/Presspass/press/2011/jul11/07-05WistronPR.mspx). Microsoft.com. 2011-07-05. . Retrieved 2012-01-05.

[120] Microsoft patent division taking cash from at least 5 Android vendors (http://www.networkworld.com/news/2011/070511-microsoft-patent-android.html).

[121] DailyTech - VIA, WTI Sell Stakes in S3 Graphics to HTC (http://www.dailytech.com/VIA+WTI+Sell+Stakes+in+S3+Graphics+to+HTC/article22078.htm).

[122] HTC to Acquire Chip Designer S3 Graphics, Securing Patents in Apple Fight - Bloomberg (http://www.bloomberg.com/news/2011-07-06/htc-to-buy-s3-graphics-securing-patents-in-apple-fight-1-.html).

[123] HTC facing backlash for S3 acquisition (http://news.cnet.com/8301-1035_3-20077720-94/htc-facing-backlash-for-s3-acquisition/?tag=mncol;txt).

[124] Microsoft wants Samsung to pay smartphone license (http://www.reuters.com/article/2011/07/06/us-samsung-microsoft-idUSTRE7651DB20110706).

[125] HTC's Flyer Tablet, Droid Accused of Infringing Apple Patents - Bloomberg (http://www.bloomberg.com/news/2011-07-11/apple-files-new-trade-complaint-against-htc-over-devices.html?cmpid=yhoo).

[126] FOSS Patents: Apple files second ITC complaint against HTC: better luck next time? (http://fosspatents.blogspot.com/2011/07/apple-files-second-itc-complaint.html).

[127] Google Acquires Over 1,000 IBM Patents in July (http://www.seobythesea.com/2011/07/google-acquires-ibm-patents-in-july/).

[128] Google's New Patents from IBM (http://www.seobythesea.com/2011/07/googles-new-patents-from-ibm/).

[129] Bad News for Android: ITC Rules HTC Violated Two Apple Patents - John Paczkowski - News - AllThingsD (http://allthingsd.com/20110715/itc-rules-htc-violated-two-apple-patents/?refcat=news).

[130] Apple Lawsuit Puts Samsung Tablet Sales in Australia on Hold (http://www.bloomberg.com/news/2011-08-01/apple-seeks-to-block-samsung-from-selling-tablet-in-australia.html).

[131] Samsung's official comment on the Australian Galaxy Tab 10.1 situation is extremely weak (http://fosspatents.blogspot.com/2011/08/samsungs-official-comment-on-australian.html).

[132] Tsukayama, Hayley (10 August 2011). "Samsung Galaxy Tab 10.1 imports banned in most of Europe" (http://www.washingtonpost.com/blogs/faster-forward/post/samsung-galaxy-tab-101-imports-banned-in-most-of-europe/2011/08/10/gIQAmnVr6I_blog.html). *The Washington Post*. . Retrieved 11 August 2011.

[133] Preliminary injunction granted by German court: Apple blocks Samsung Galaxy Tab 10.1 in the entire European Union except for the Netherlands (http://fosspatents.blogspot.com/2011/08/preliminary-injunction-granted-by.html).

[134] "Microsoft files to ban imports of Motorola smartphones" (http://androidandme.com/2011/08/news/microsoft-files-to-ban-imports-of-motorola-smartphones). Android and Me. . Retrieved 2011-12-10.

[135] Womack, Brian (2011-08-22). "Motorola Value Found in 18 Patents Used Against Apple: Tech" (http://www.bloomberg.com/news/2011-08-22/motorola-s-value-for-google-found-in-18-patents-used-against-apple-tech.html). Bloomberg. . Retrieved 2011-12-10.

[136] Leske, Nicola (2011-08-16). "Samsung Galaxy tablet ban lifted in most of Europe" (http://uk.reuters.com/article/2011/08/16/tech-us-samsung-apple-ban-idUKTRE77F45420110816). *Reuters*. . Retrieved 16 August 2011.

[137] "Samsung Galaxy Tab ban is on hold" (http://www.bbc.co.uk/news/business-14548895). *BBC*. 2011-08-16. . Retrieved 16 August 2011.

[138] Google and IBM do it again: Google Acquires over 1,000 Patents from IBM in August (http://www.seobythesea.com/2011/09/google-ibm-patents-august/#more-6675).

[139] Microsoft Asks For An Import Ban On Motorola Smartphones (http://techcrunch.com/2011/08/23/microsoft-asks-for-an-import-ban-on-motorola-smartphones/).

[140] "Microsoft files to ban imports of Motorola smartphones" (http://androidandme.com/2011/08/news/microsoft-files-to-ban-imports-of-motorola-smartphones/). Android and Me. . Retrieved 2011-12-10.
[141] Euro ban for Samsung Galaxy phone (http://www.bbc.co.uk/news/technology-14652482).
[142] Samsung Puts Galaxy 10.1 Tablet on Hold as Apple Wins German Court Order (http://www.bloomberg.com/news/2011-09-04/apple-wins-german-injunction-on-new-galaxy-tab.html).
[143] FOSS Patents: Apple to ITC: Andy Rubin got inspiration for Android framework while working at Apple, hence infringes an Apple API patent (http://fosspatents.blogspot.com/2011/09/apple-to-itc-andy-rubin-got-inspiration.html).
[144] Milford, Phil (2011-09-08). "HTC Sues Apple Using Google Patents Bought Last Week as Battle Escalates" (http://www.bloomberg.com/news/2011-09-07/htc-sues-apple-alleging-infringement-of-four-u-s-patents.html). Bloomberg. . Retrieved 2011-12-10.
[145] Google Hands HTC Patents to Use Against Apple in Smartphone Wars (http://www.businessweek.com/news/2011-09-08/google-hands-htc-patents-to-use-against-apple-in-smartphone-wars.html).
[146] Patel, Nilay (2011-09-07). "HTC sues Apple for patent infringement... using patents purchased by Google" (http://thisismynext.com/2011/09/07/htc-sues-apple-patent-infringement-patents-purchased-google/). Thisismynext.com. . Retrieved 2012-01-05.
[147] Google gets its hands dirty - Apple 2.0 - Fortune Tech (http://tech.fortune.cnn.com/2011/09/08/google-gets-its-hands-dirty/).
[148] "Microsoft and Acer Sign Patent License Agreement: Agreement will cover Acer's Android tablets and smartphones" (http://www.microsoft.com/Presspass/press/2011/sep11/09-08AcerPR.mspx). Microsoft.com. 2011-09-08. . Retrieved 2012-01-05.
[149] "Microsoft and ViewSonic Sign Patent Agreement: Agreement will cover ViewSonic's Android Tablets and smartphones" (http://www.microsoft.com/Presspass/press/2011/sep11/09-08ViewSonicPR.mspx). Microsoft.com. 2011-09-08. . Retrieved 2012-01-05.
[150] "Microsoft ropes two more OEMs into Android patent deal" (http://www.zdnet.com/blog/hardware/microsoft-ropes-two-more-oems-into-android-patent-deal/14622). Zdnet.com. . Retrieved 2012-01-05.
[151] Tu, Janet I. (2011-09-08). "Microsoft signs Android patent agreements with Acer, ViewSonic" (http://seattletimes.nwsource.com/html/microsoftpri0/2016144500_microsoft_signs_android_patent_agreements_with_ace.html). Seattletimes.nwsource.com. . Retrieved 2012-01-05.
[152] Matussek, Karin (2011-09-09). "Apple Wins German Ban on Samsung Tablet" (http://www.bloomberg.com/news/2011-09-09/apple-wins-ruling-for-german-samsung-galaxy-tablet-10-1-ban.html). Bloomberg.com. . Retrieved 2012-01-05.
[153] "Latest Samsung lawsuit targets Apple's iPhone, iPad in France" (http://www.appleinsider.com/articles/11/09/13/latest_samsung_lawsuit_targets_apples_iphone_ipad_in_france.html). AppleInsider. 2011-09-13. . Retrieved 2012-01-05.
[154] Meyer, David (2011-09-14). "Apple sues Samsung in the UK over Android | ZDNet UK" (http://www.zdnet.co.uk/blogs/communication-breakdown-10000030/apple-sues-samsung-in-the-uk-over-android-10024342/). Zdnet.co.uk. . Retrieved 2012-01-05.
[155] "USPTO Assignments on the Web" (http://assignments.uspto.gov/assignments/q?db=pat&reel=026894&frame=0001). Assignments.uspto.gov. . Retrieved 2012-01-05.
[156] "Samsung fires back at Apple in Australia with countersuit against iPhone, iPad" (http://www.appleinsider.com/articles/11/09/16/samsung_files_patent_case_against_apple_in_australia_over_iphone_ipad.html). Appleinsider.com. . Retrieved 2012-01-05.
[157] "Microsoft and Samsung Broaden Smartphone Partnership: Agreements mark new initiatives to promote Windows Phone and share intellectual property" (http://www.microsoft.com/Presspass/press/2011/sep11/09-28SamsungPR.mspx). Microsoft.com. 2011-09-28. . Retrieved 2012-01-05.
[158] Our Licensing Deal with Samsung: How IP Drives Innovation and Collaboration - Microsoft on the Issues - Site Home - TechNet Blogs (http://blogs.technet.com/b/microsoft_on_the_issues/archive/2011/09/28/our-licensing-deal-with-samsung-how-ip-drives-innovation-and-collaboration.aspx).
[159] Florian Mueller (2011-09-28). "FOSS Patents: Samsung takes Android patent license from Microsoft rather than wait for Motorola" (http://fosspatents.blogspot.com/2011/09/samsung-takes-android-patent-license.html). Fosspatents.blogspot.com. . Retrieved 2012-01-05.
[160] "Apple Wins Injunction Blocking Sale of Galaxy Tab 10.1 in Australia" (http://www.macrumors.com/2011/10/12/apple-wins-injunction-blocking-sale-of-galaxy-tab-10-1-in-australia/). Mac Rumors. 2011-10-12. . Retrieved 2012-01-05.
[161] "Microsoft and Quanta Computer Sign Patent Agreement Covering Android and Chrome-Based Devices - Redmond, Wash., Oct. 13, 2011 /PRNewswire/" (http://www.prnewswire.com/news-releases/microsoft-and-quanta-computer-sign-patent-agreement-covering-android-and-chrome-based-devices-131793888.html). Washington: Prnewswire.com. . Retrieved 2012-01-05.
[162] Microsoft Inks Another Android Patent Deal, This Time With Quanta | TechCrunch (http://techcrunch.com/2011/10/13/microsoft-inks-another-android-patent-deal-this-time-with-quanta/).
[163] Levine, Dan (2011-10-13). "U.S. judge says Samsung tablets infringe Apple patents" (http://www.reuters.com/article/2011/10/13/us-apple-samsung-lawsuit-idUSTRE79C79C20111013?feedType=RSS&feedName=businessNews&utm_source=dlvr.it&utm_medium=twitter&dlvrit=56943). Reuters.com. . Retrieved 2012-01-05.
[164] Apple Inc.. "Distribute your App - iOS Developer Program - Apple Developer" (http://developer.apple.com/programs/ios/distribute.html). Developer.apple.com. . Retrieved 2012-01-05.
[165] "Apple's rivals battle for iOS scraps as app market sales grow to $2.2 billion" (http://www.appleinsider.com/articles/11/02/18/rim_nokia_and_googles_android_battle_for_apples_ios_scraps_as_app_market_sales_grow_to_2_2_billion.html). Appleinsider.com. 2011-02-18. . Retrieved 2012-01-05.
[166] "Google Android has double the number of free apps than Apple's App Store" (http://techcrunch.com/2010/07/05/distimo-june-2010/). Distimo. 15 July 2009. . Retrieved 15 July 2009.

[167] "McAfee: Android malware surges 76%, iPhone untouched" (http://www.electronista.com/articles/11/08/23/mcafee.shows.android.facing.huge.spike.in.malware/). Electronista.com. . Retrieved 2012-01-05.
[168] Dalrymple, Jim (2011-11-16). "Android sees a 472% increase in malware since July" (http://www.loopinsight.com/2011/11/16/android-sees-a-472-increase-in-malware-since-july/). Loopinsight.com. . Retrieved 2012-01-05.
[169] Mobile Malware Development Continues To Rise, Android Leads The Way (http://globalthreatcenter.com/?p=2492).
[170] "The Mother Of All Android Malware Has Arrived" (http://www.androidpolice.com/2011/03/01/the-mother-of-all-android-malware-has-arrived-stolen-apps-released-to-the-market-that-root-your-phone-steal-your-data-and-open-backdoor/). *Android Police*. March 6, 2011. .
[171] Perez, Sarah (2009-02-12). "Android Vulnerability So Dangerous, Owners Warned Not to Use Phone's Web Browser" (http://www.readwriteweb.com/archives/android_vulnerability_so_dangerous_shouldnt_use_web_browser.php). Readwriteweb.com. . Retrieved 2011-08-08.
[172] "Lookout, Retrevo warn of growing Android malware epidemic, note Apple's iOS is far safer" (http://www.appleinsider.com/articles/11/08/03/lookout_retrevio_warn_of_growing_android_malware_epidemic_note_apples_ios_is_far_safer.html). Appleinsider.com. 2011-08-03. . Retrieved 2012-01-05.
[173] "First SMS Trojan detected for smartphones running Android" (http://www.kaspersky.com/news?id=207576158). Kaspersky Lab. . Retrieved 2010-10-18.
[174] "Apple's iOS unaffected by malware as Android exploits surge 76%" (http://www.appleinsider.com/articles/11/08/24/apples_ios_unaffected_by_malware_as_android_exploits_surge_76.html). Appleinsider.com. 2011-08-24. . Retrieved 2012-01-05.
[175] "Security researcher finds code signing flaw that opens door for iOS malware" (http://www.appleinsider.com/articles/11/11/07/newly_found_code_signing_flaw_allows_for_ios_malware.html). Appleinsider.com. 2011-11-07. . Retrieved 2012-01-05.
[176] Apple's iOS more secure than Google's Android, says Symantec (http://www.appleinsider.com/articles/11/06/28/apples_ios_more_secure_than_googles_android_says_symantec.html)
[177] the understatement: Android Orphans: Visualizing a Sad History of Support (http://theunderstatement.com/post/11982112928/android-orphans-visualizing-a-sad-history-of-support)
[178] "Smart phones: how to stay clever in downturn" (http://www.deloitte.co.uk/TMTPredictions/telecommunications/Smartphones-clever-in-downturn.cfm). *Deloitte Telecommunications Predictions*. .
[179] "Android Phones Steal Market Share" (http://bmighty.informationweek.com/mobile/showArticle.jhtml?articleID=224201881). .
[180] "100 Million Club – H1 2010" (http://www.visionmobile.com/blog/2010/10/smart-feature-phones-the-unbalanced-equation-100-million-club-series/). .
[181] "Mobile Phone Development » Blog Archive » Gartner Q3 2010" (http://www.mobilephonedevelopment.com/archives/1149). Mobilephonedevelopment.com. 2010-11-10. . Retrieved 2011-09-07.
[182] Stan Schroeder (2011-02-10). "Gartner: Symbian Is Still the Number One Smartphone Platform" (http://mashable.com/2011/02/10/symbian-number-one-gartner-report/). Mashable.com. . Retrieved 2011-09-07.
[183] "Gartner Says Worldwide Mobile Device Sales to End Users Reached 1.6 Billion Units in 2010; Smartphone Sales Grew 72 Percent in 2010" (http://www.gartner.com/it/page.jsp?id=1543014). Gartner.com. 2011-02-11. . Retrieved 2011-09-07.
[184] "Olswang Predicts Mobile's Impact on TV: The Convergent Revolution" (http://www.tvgenius.net/blog/2011/03/17/mobile-tv-convergence/). Tvgenius.net. 2011-03-17. . Retrieved 2011-09-07.
[185] "Berg: Smartphone shipments grew 74% in 2010" (http://www.bgr.com/2011/03/10/berg-smartphone-shipments-grew-74-in-2010/). *Boy Genius Report*. March 10, 2011. .
[186] Generation App: 62% of Mobile Users 25-34 own Smartphones | Nielsen Wire (http://blog.nielsen.com/nielsenwire/?p=29786).
[187] Market Research | Consumer Market Research - NPD - As Smartphone Prices Fall, Retailers Are Leaving Money on the Table, According to The NPD Group (http://www.npdgroup.com/wps/portal/npd/us/news/pressreleases/pr_111114a)
[188] Apple, With 4 Percent of Handset Market, Captures 52 Percent of Profits (http://www.pcmag.com/article2/0,2817,2395951,00.asp#fbid=SLgcdJun7IU)
[189] "Smartphones killing point-and-shoots, now take almost 1/3 of photos" (http://gigaom.com/2011/12/22/smartphones-killing-point-and-shoots-now-take-almost-13-of-photos/). . Retrieved December 25, 2011.
[190] Tarmo Virki (2011-01-31). "Google topples Symbian from smart phones top spot" (http://www.theglobeandmail.com/news/technology/mobile-technology/google-topples-symbian-from-smartphones-top-spot/article1888557/?cmpid=rss1). Toronto: Theglobeandmail.com. . Retrieved 2011-09-07.
[191] Virki, Tarmo (2011-05-04). "Android became clear smartphone leader in first quarter: Canalys" (http://www.reuters.com/article/2011/05/04/us-smartphones-research-idUSTRE7434VV20110504). Reuters.com. . Retrieved 2011-09-07.
[192] "Android Takes Over in UK" (http://www.bestsmartphone.com/2011/11/02/android-takes-over-in-uk/). bestsmartphone.com. 2011-11-02. .
[193] http://www.gartner.com/it/page.jsp?id=910112
[194] http://www.gartner.com/it/page.jsp?id=910112
[195] http://www.gartner.com/it/page.jsp?id=1306513
[196] http://www.gartner.com/it/page.jsp?id=1543014
[197] iPass : Mobile Workforce Report (http://mobile-workforce-project.ipass.com/).

[198] iPhone pushes past BlackBerry to top enterprise phone ranks (http://www.appleinsider.com/articles/11/11/16/iphone_pushes_past_blackberry_to_top_enterprise_phone_ranks.html)
[199] In the US Market, iPhone Outperforms Other Mobile Platforms in User Loyalty by a Wide Margin, Android is Second, Blackberry Fourth (http://www.zokem.com/2011/01/in-the-us-market-iphone-outperforms-other-mobile-platforms-in-user-loyalty-by-a-wide-margin-android-is-second-blackberry-fourth/)
[200] iPhone Whips Android, BlackBerry in User Loyalty: Zokem (http://www.eweek.com/c/a/Mobile-and-Wireless/iPhone-Whips-Android-Blackberry-in-User-Loyalty-Zokem-445681/)
[201] Strategy Analytics: Samsung Becomes World's Number One Smartphone Vendor in Q3 2011 - MarketWatch (http://www.marketwatch.com/story/strategy-analytics-samsung-becomes-worlds-number-one-smartphone-vendor-in-q3-2011-2011-10-27)
[202] http://www.gartner.com/it/page.jsp?id=1764714
[203] "Smartphone Sales Will Hit 420 Million In 2011, To Take 28 Percent Of The Total Phone Market" (http://techcrunch.com/2011/07/27/smartphone-sales-will-hit-420-million-in-2011-to-take-28-percent-of-the-total-phone-market/). TechCrunch. 27 July 2011. .
[204] Apple's iPhone accounted for 66% of Q2 smartphone profit among top vendors (http://www.bgr.com/2011/07/29/apples-iphone-accounted-for-66-of-q2-smartphone-profit-among-top-vendors/)
[205] "SA agrees: Apple now top smartphone vendor in the world with 140% growth" (http://www.bgr.com/2011/07/29/sa-agrees-apple-now-top-smartphone-vendor-in-the-world-with-240-growth/). BGR. 29 July 2011. .
[206] "Nokia reports Q2 results: Sells 88.5 million devices and declares net loss of 368 million euros" (http://mobilesyrup.com/2011/07/21/nokia-reports-q2-results-sells-88-5-million-devices-and-declares-net-loss-of-368-million-euros/). Mobilesyrup.com. 2011-07-21. . Retrieved 2011-09-07.
[207] "In U.S. Smartphone Market, Android is Top Operating System, Apple is Top Manufacturer" (http://blog.nielsen.com/nielsenwire/?p=28516). Nielsen. 28 July 2011. . Retrieved 5 September 2011.
[208] Isaac, Mike (28 July 2011). "Android Still Dominates Phones, But What About the Rest of Mobile?" (http://www.wired.com/gadgetlab/2011/07/android-ios-platform-share/all/1). Wired. . Retrieved 5 September 2011.
[209] iPhone 4 remains top-selling US smartphone despite growing iPhone 5 hype (http://www.appleinsider.com/articles/11/09/06/iphone_4_remains_top_selling_us_smartphone_despite_growing_iphone_5_hype.html)
[210] Previous-gen Apple iPad, iPhone 3GS often outsell new Android devices (http://www.appleinsider.com/articles/11/05/09/previous_gen_apple_ipad_iphone_3gs_often_outsell_new_android_devices.html)
[211] "Samsung Becomes Biggest Smartphone Vendor, as Android's Market Share Grows" (http://www.pcworld.com/article/243861/samsung_becomes_biggest_smartphone_vendor_as_androids_market_share_grows.html). PCWorld. 2011-11-15. . Retrieved 2011-12-10.
[212] Ever-Popular iPhone Named Top Smartphone. Again (http://www.wired.com/gadgetlab/2011/09/iphone-tops-survey-sixth-time/all/1).
[213] Wireless Consumer Smartphone Ratings (Volume 1) (http://www.jdpower.com/Electronics/ratings/wireless-consumer-smartphone-ratings-(volume-1)/)
[214] Wireless Consumer Smartphone Ratings (Volume 2) (http://www.jdpower.com/Electronics/ratings/wireless-consumer-smartphone-ratings-(volume-2)/).
[215] 2011 U.S. Wireless Handset Customer Satisfaction Studies—Vol. 2 (http://www.jdpower.com/news/pressRelease.aspx?ID=2011146)
[216] 2011 Wireless Smartphone and Traditional Mobile Phone Satisfaction Studies-Vol. 1 (http://www.jdpower.com/news/pressrelease.aspx?ID=2011030)
[217] 2010 U.S. Wireless Smartphone and Traditional Mobile Phone Customer Satisfaction Studies-Volume 1 (http://businesscenter.jdpower.com/news/pressrelease.aspx?ID=2010039)
[218] 2009 Wireless Consumer Smartphone and Traditional Mobile Phone Customer Satisfaction Studies (http://businesscenter.jdpower.com/news/pressrelease.aspx?ID=2009082)
[219] Menezes, Gary (2010-09-11). "Symbian OS, Now Fully Open Source" (http://www.watblog.com/2010/02/06/symbian-os-now-fully-open-source). Watblog.com. . Retrieved 2011-09-07.
[220] "Nokia N9 Smart Phone Review" (http://wontek.com/smart-phones/nokia/nokia-n9-smart-phone-review). Wontek.com. 2011-01-01. . Retrieved 2011-09-07.
[221] Lance Whitney, CNET. " Nearly 1 in 5 smartphone owners use check-in services (http://news.cnet.com/8301-1023_3-20062640-93.html?part=rss&subj=news&tag=2547-1_3-0-20)." May 13, 2011. Retrieved May 13, 2011.
[222] Second Screen change the way we watch TV (http://www.good.is/post/double-the-glow-will-second-screen-apps-change-the-way-we-watch-tv/)
[223] explosion of second screen apps (http://www.adweek.com/news/technology/second-screen-apps-explode-132237)

External links

- Freesmartphone.org (http://www.freesmartphone.org) – a collaboration platform for open-source smartphones
- Defining the Smartphone, part 1 (http://www.allaboutsymbian.com/features/item/Defining_the_Smartphone.php), part 2 (http://www.allaboutsymbian.com/features/item/Spy_versus_Spy_No_Smartphone_versus_Smartphone.php) by Steve Litchfield on July 16, 2010, various definitions are treated
- Mozilla Labs: Seabird –Community driven Mobile Phone Concept (http://mozillalabs.com/conceptseries/2010/09/23/seabird/)

Visto

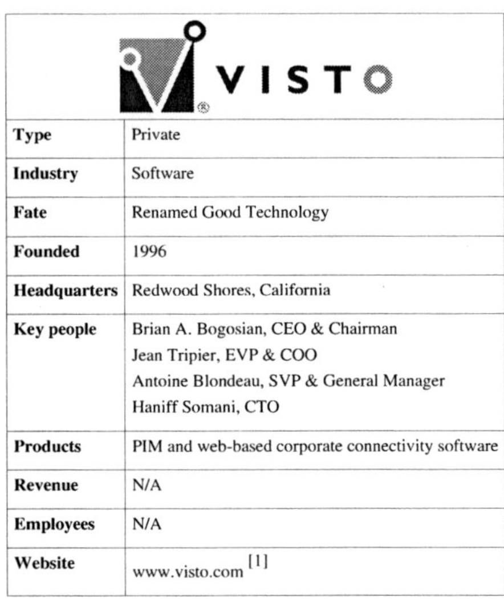

Type	Private
Industry	Software
Fate	Renamed Good Technology
Founded	1996
Headquarters	Redwood Shores, California
Key people	Brian A. Bogosian, CEO & Chairman Jean Tripier, EVP & COO Antoine Blondeau, SVP & General Manager Haniff Somani, CTO
Products	PIM and web-based corporate connectivity software
Revenue	N/A
Employees	N/A
Website	www.visto.com [1]

Visto was a private US company located in Redwood Shores, CA. Visto sold Push e-mail products for mobile phones. The company acquired Good Technology in 2009 and renamed itself for the acquired company.

Visto the Company

Visto was founded as "Roampage" in 1996 by ex-Javasoft employees Chris Zuleeg, Prasad Wagle, and Hong Bui. Doug Brackbill was brought on as CEO. David Cowan was one of the initial investors. By late 1997 there were over 40 employees, and the company name was changed to Visto.

In 1996, the target customer was the mobile professional. When this user base was not expanding as fast as desired, a free Visto Beta service was opened to anyone by summer 1997, with the first "Commercial Release" in April 1998, when functionality and services above the baseline were offered for a fee. Customer growth increased, but never generated profitable revenue stream. Consumer services were terminated circa 2002, when Visto returned focus to OEM and corporate customers.

In 1998, Visto was in talks with Palm Computing, and Wireless Knowledge (a Qualcomm-Microsoft joint venture). Visto partnered with Geocities before Geocities was acquired by Yahoo.

In 2003, Visto bought push email solutions business from Psion, which was created in 2002 as a new division called Psion Software.

On February 23, 2009 Visto announced it is buying Good Technology from Motorola, Visto closed acquisition of Good Technology on March 2, 2009 and changed the Visto name to Good Technology.

Visto the Product/Service

In 1996, Visto offered what was likely the first complete browser-based application suite: including email, to-do list, calendar, address book, remote file storage (ftp by any other name), browser bookmarks (both IE and Netscape). Visto pioneered the sharing of calendars between customers, allowing one to send invites and coordinate events with other Visto users.

Visto also offered a then-unique push email service allowing corporate email to be accessed outside the company firewall. A Windows service was configured on a machine within the company firewall, which pushed email to the easily-accessed Visto servers.

The complete application suite was synchronize-able between Windows-based machines equipped with the Visto software. Syncing to personal information managers (PIMs)--such as Microsoft Outlook, Lotus Notes, Lotus Organizer, and ACT!--was achieved by licensing IntelliSync technology. Synchronization was limited to Windows machines, but the synchronized information was available from any machine running Internet Explorer or Netscape Navigator/Communicator. Other browsers were not supported.

Initial code development was in Java, but performance proved too slow for customer satisfaction and scalability. In late 1997, early 1998, the transition of the complete code base from Java to C++ cost the company six months of product advancement, diminishing market lead and allowing competitors such as Yahoo to catch up with the functionality Visto offered for their own consumer services.

In 2008, Visto provides only corporate email services, billing itself as a "mobile email provider."

Patent litigation

In December 2005 Visto filed a lawsuit against Microsoft for "misuse of proprietary technology".[2] Also in December 2005, Visto signed a licensing agreement with NTP, Inc., further increasing its patent pool, in exchange for NTP receiving an equity stake in Visto. In February 2006, Visto sued competitor Good Technology for alleged violations of four patents related to Good's wireless email technology. Again on May 1, 2006, Visto sued competitor, Research In Motion for alleged patent infringement.[3]

On 3 March 2008, Visto and Microsoft announced they had reached a settlement and licensing agreement involving cash and non-cash consideration. Both companies will dismiss all pending legal claims.[4]

In July 2009, the case was settled, with Research in Motion paying Visto $267.5 million.[5]

See also

- Good Technology
- Palm
- Consilient
- Psion
- Patent troll

References

[1] http://www.visto.com/
[2] "Visto Files Legal Action Against Microsoft for Misuse of Visto's Proprietary Technology" (http://www.prnewswire.com/cgi-bin/stories.pl?ACCT=104&STORY=/www/story/12-15-2005/0004234488&EDATE=). PRNewswire. 2005-12-15. . Retrieved 2008-03-03.
[3] (http://ipcenter.bna.com/pic2/ip.nsf/id/BNAP-6PNTRU?OpenDocument)
[4] "Visto says ends patent dispute with Microsoft" (http://news.yahoo.com/s/nm/20080303/tc_nm/visto_microsoft_dc_1). Yahoo! News/Reuters. 2008-03-03. . Retrieved 2008-03-03.
[5] Silver, Sara; Weinberg, Stuart (2009-07-17). "RIM Settles Patent Battle With Visto" (http://online.wsj.com/article/SB124775322296051701.html). *The Wall Street Journal*. .

External links

- Official website (http://www.visto.com/)
- Who has time for this - David Cowan (Visto Founder)'s Blog (http://whohastimeforthis.blogspot.com)

Secure_Digital

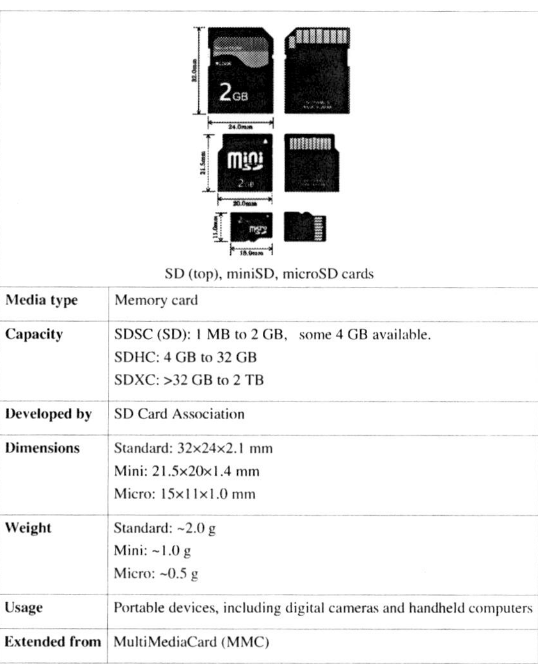

SD (top), miniSD, microSD cards

Media type	Memory card
Capacity	SDSC (SD): 1 MB to 2 GB, some 4 GB available. SDHC: 4 GB to 32 GB SDXC: >32 GB to 2 TB
Developed by	SD Card Association
Dimensions	Standard: 32×24×2.1 mm Mini: 21.5×20×1.4 mm Micro: 15×11×1.0 mm
Weight	Standard: ~2.0 g Mini: ~1.0 g Micro: ~0.5 g
Usage	Portable devices, including digital cameras and handheld computers
Extended from	MultiMediaCard (MMC)

Secure Digital (**SD**) is a non-volatile memory card format developed by the SD Card Association (SDA) for use in portable devices. The SD technology is used by more than 400 brands across dozens of product categories and more than 8,000 models.[1]

SD comprises several families of cards:[2] the original, Standard-Capacity (SDSC) card, a High-Capacity (SDHC) card family, an eXtended-Capacity (SDXC) card family,[3] and the SDIO[4] family with input/output functions rather than just data storage.

SD also comprises three different form factors: the original size, the "mini" size, and the "micro" size (see illustration). Electrically passive adaptors allow the use of a smaller card in a host device built to hold a larger card. There are many combinations of form factors and device families.

Host devices that comply with newer versions of the specification provide backward compatibility and accept older SD cards, but older host devices do not recognize newer cards. The SDA uses several trademarked logos to enforce compliance with its specifications and assure users of compatibility.[5] This article explains several factors that can prevent the use of a newer SD card:

- A newer card may offer greater capacity than the host device can handle.
- A newer card may use a file system the host device cannot navigate.
- Use of an SDIO card requires the host device be designed for the input/output functions the card provides.
- The organization of the card was changed starting with the SDHC family.
- Some vendors produced SDSC cards above 1 GB before the SDA had standardized a method of doing so.

Types of cards

The SDA extended the SD specification in various ways:

- It defined electrically identical cards in smaller sizes: miniSD and microSD (originally named TransFlash or TF). Smaller cards are usable in larger slots through use of a passive adapter. By comparison, Reduced Size MultiMediaCards (RS-MMCs) are simply shorter MMCs and can be used in MMC slots by use of a physical extender.
- It defined higher-capacity cards, some with faster speeds and added capabilities: SDHC (Secure Digital High Capacity) and SDXC (Secure Digital eXtended Capacity). These cards redefine the interface so that they cannot be used in older host devices.
- It defined an SDIO card family that provides input-output functions and may also provide memory functions. These cards are only fully functional in host devices designed to support their input-output functions.

Physical size

The SD card specification defines three physical sizes. The SD and SDHC families are available in all three sizes, but the SDXC family is not available in the mini size, and the SDIO family is not available in the micro size.

Size comparison of families: SD, miniSD, microSD

Standard size

- SD (SDSC), SDHC, SDXC, SDIO
- 32 mm × 24 mm × 2.1 mm
- 32 mm × 24 mm × 1.4 mm (as thin as MMC) for **Thin SD** (rare)

Mini size

- miniSD, miniSDHC, miniSDIO
- 21.5 mm × 20 mm × 1.4 mm

Micro size

The microSD form factor is the smallest memory card format currently available.

- microSD, microSDHC, microSDXC
- 15 mm × 11 mm × 1.0 mm

SDHC

16GB SDHC Class 6

The Secure Digital High Capacity (SDHC) format, defined in Version 2.0 of the SD specification, supports cards with capacities up to 32 GB.[1] The SDHC trademark is licensed to ensure compatibility.[6]

SDHC cards are physically and electrically identical to standard-capacity SD cards (SDSC). The major compatibility issues between SDHC and SDSC cards are the redefinition of the Card-Specific Data (CSD) register in Version 2.0 (see below), and the fact that SDHC cards are shipped preformatted with the FAT32 file system.

Host devices that accept SDHC cards are required to accept SDSC cards.[1] However, host devices designed for SDSC do not recognize SDHC or SDXC memory cards, although some devices can do so through a firmware upgrade.[7] Older operating systems require patches to support SDHC. For instance, Microsoft Windows XP before SP3 requires a patch to support access to SDHC cards.[8] Windows Vista SP1 also requires a later service pack.[9] [10]

SDXC

The Secure Digital Extended Capacity (SDXC) format supports cards up to 2 TB (2048 GB), compared to a limit of 32 GB for SDHC cards in the SD 2.0 specification.

History

SDXC was announced at Consumer Electronics Show (CES) 2009 (January 7–10, 2009). At the same show, SanDisk and Sony also announced a comparable Memory Stick XC variant with the same 2 TB maximum as SDXC,[11] and Panasonic announced plans to produce 64 GB SDXC cards.[12]

On March 6, 2009, Pretec introduced the first SDXC card,[13] a 32 GB card with a read/write speed of 400 Mbit/s. But it was not until early 2010 that compatible host devices came onto the market, including Sony's Handycam HDR-CX55V camcorder, Canon's EOS Rebel T2i Digital SLR camera,[14] a USB card reader from Panasonic, and an integrated SDXC card reader from JMicron.[15] The earliest laptops to integrate SDXC card readers relied on a USB 2.0 bus, which does not have the bandwidth to support SDXC at full speed.[16]

Also in early 2010, commercial SDXC cards appeared from Toshiba (64 GB),[17] [18] Panasonic (64 GB and 48 GB),[19] and SanDisk (64 GB).[20] In early 2011, Centon Electronics, Inc. (64 GB and 128 GB) and Lexar (128 GB)[21] began shipping SDXC cards rated at Speed Class 10. Pretec offered cards from 8 GB to 128 GB rated at Speed Class 16.[22]

In September 2011, SanDisk released a 64 GB microSDXC card.[23] Kingmax released a comparable product in 2011.[24]

Compatibility with SDHC

SDXC host devices accept all previous families of SD memory cards.[25] Conversely, SDHC host devices will accept SDXC cards that follow Version 3.0, since the interface is identical,[3] but the following issues may affect usability:

- SDXC cards are pre-formatted with Microsoft's proprietary and patented exFAT file system, which the host device might not support. Since Microsoft does not publish the specifications of exFAT and its use requires a non-free license, many alternative or older operating systems do not support exFAT for technical or legal reasons. The use of exFAT on some SDXC cards may render SDXC unsuitable as a universal exchange medium, as an SDXC card that uses exFAT would not be usable in all host devices. Since the FAT32 file system supports volumes up to the SDXC's maximum theoretical capacity of 2 TB as well, a user could reformat an SDXC card to use FAT32 for greater portability between computers (see below), but consumer products such as digital cameras and camcorders do not normally provide the choice to format SDXC cards as FAT32 nor do they accept FAT32-reformatted SDXC cards.

- An SDHC host device will not test the new capability bits defined for SDXC cards. It will therefore not be able to use the new features of SDXC, such as transfer speeds above UHS104.[26]

Host Operating System support

Microsoft Windows versions that support SDXC are: Windows 7, Windows Vista SP1+,[3] Windows XP SP2 or SP3 with KB955704,[27] Windows Server 2008 SP1+, Windows Server 2003 SP2 or SP3 with KB955704, and Windows CE 6.0 and higher.

Apple Mac OS X versions that support SDXC cards and exFAT are Mac OS X Snow Leopard 10.6.5 or later including OS X Lion 10.7.[28] [29]

BSD and Linux systems that support SDHC cards also support SDXC cards that contain a compatible file system, but they normally do not support the proprietary exFAT file system that is installed on SDXC cards, because of patent issues. The user may reformat the card to contain a different file system (see below).

SDIO

A SDIO (Secure Digital Input Output) card is an extension of the SD specification to cover I/O functions. Host devices that support SDIO (typically PDAs like the Palm Treo, but occasionally laptops or mobile phones) can use the SD slot to support GPS receivers, modems, barcode readers, FM radio tuners, TV tuners, RFID readers, digital cameras, and interfaces to Wi-Fi, Bluetooth, Ethernet, and IrDA. Many other SDIO devices have been proposed, but it is now more common for I/O devices to connect using the USB interface.

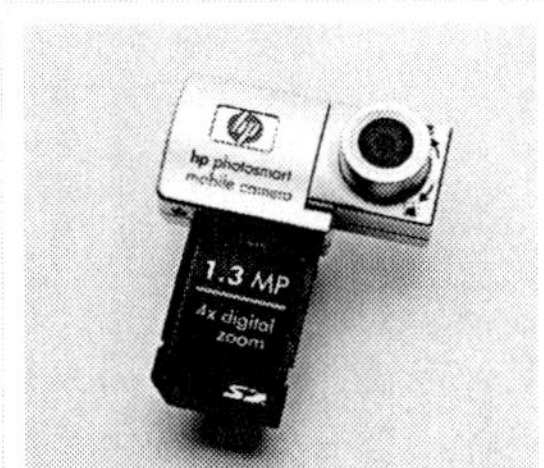

Camera using the SDIO interface to connect to some HP iPAQ devices

SDIO cards support most of the memory commands of SD cards. SDIO cards can be structured as 8 logical cards, although currently, the typical way that an SDIO card uses this capability is to structure itself as one I/O card and one memory card.

Host support for SDIO

The SDIO and SD interfaces are mechanically and electrically identical. Host devices built for SDIO cards generally accept SD memory cards without I/O functions. However, the reverse is not true, because host devices need suitable drivers and applications to support the card's I/O functions. For example, an HP SDIO camera usually does not work with PDAs that do not list it as an accessory. Inserting an SDIO card into any SD slot causes no physical damage nor disruption to the host device, but users may be frustrated that the SDIO card does not function fully when inserted into a seemingly compatible slot. (Bluetooth devices exhibit comparable compatibility issues, although to a lesser extent thanks to standardized Bluetooth profiles.)

Features

Card security (non-DRM)

Write protection

When looking at the card from the top, the right side (the side with the beveled corner) must be notched.

On the left side, there may be a write-protection notch. If the notch is omitted, the card can be read and written. If the card is notched, it is read-only, except that the notch may be partly covered by a sliding tab. In this case, the user can slide the tab upward (toward the contacts) to declare the card read/write, or downward to declare it read-only.

The presence of a notch, and the presence and position of a tab, have no effect on the SD card's operation. A host device that supports write protection should refuse to write to an SD card that is designated read-only in this way. Some host devices do not support write protection, which is an optional feature of the SD specification. Host devices that do obey a read-only indication may give the user a way to override it. The miniSD and microSD formats do not support write protection with the notch-and-tab method.

Typical cards sold as a medium for protected content are permanently marked read-only by having a notch and no sliding tab.

Card password

A host device can lock an SD card using a password of up to 16 bytes, typically supplied by the user. A locked card interacts normally with the host device except that it rejects commands to read and write data. A locked card can be unlocked only by providing the same password. The host device can, after supplying the old password, specify a new password or disable locking. Without the password (typically, in the case that the user forgets the password), the host device can command the card to erase all the data on the card for future re-use (except card data under DRM), but there is no way to gain access to the existing data.

DRM features

All SD cards incorporate a digital rights management (DRM) scheme. Roughly 10% of the storage capacity of an SD card is not available to the user, but is used by the on-card processor to verify the identity of an application program that it will then allow to read protected content. The card prohibits other accesses, such as users trying to make copies of protected files.

The DRM scheme embedded in the SD cards is the Content Protection for Recordable Media (CPRM or CPPM) specification of the 4C Entity, which features the Cryptomeria cipher (also termed *C2*). The specification is kept secret and is accessible only to licensees. The scheme has not been broken or hacked, but this feature of SD cards is rarely used to protect content. DVD-Audio uses the same DRM scheme.

Windows Media files can be DRM-encoded so as to make use of the SD card's DRM abilities.

Windows Phone 7 devices use SD cards designed to be accessed only by the phone manufacturer or mobile provider. An SD card inserted into the phone underneath the battery compartment becomes locked "to the phone with an automatically generated key" so that "the SD card cannot be read by another phone, device, or PC".[30] Symbian devices, however, are some of the very few which can perform the necessary low-level format operations on locked SD cards. It is therefore possible to use a device such as the Nokia N8 to reformat the card for subsequent use in other devices.[31]

The *Super Digital* cards manufactured by Super*Talent are the same in appearance and function as Secure Digital cards, but they lack the CPRM feature of Secure Digital cards.[32]

Vendor enhancements

Vendors have sought to differentiate their products in the market through various vendor-specific features:

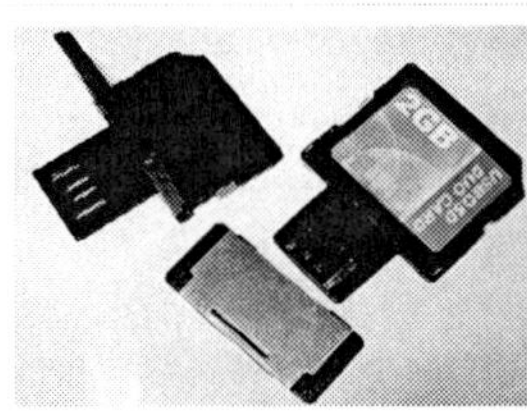
SD cards with dual-interface SD and USB connections

- **Integrated Wi-Fi** — Eye-Fi produces several SD cards with built-in Wi-Fi transceivers providing static security (WEP 40; 104; and 128, WPA-PSK, and WPA2-PSK). The card lets any digital camera with an SD slot transmit captured images over a wireless network, or store the images on the card's memory until it is in range of a wireless network.[33] Some models geotag their pictures.
- **Pre-loaded content** — In 2006, SanDisk announced Gruvi, a microSD card with extra Digital Rights Management (DRM) features, with which to use memory cards as a medium to sell content. SanDisk again announced pre-loaded cards in 2008, under the slotMusic name, this time not using any of the DRM capabilities of the SD card.[34] In 2011, SanDisk offered various collections of 1000 songs on a single slotMusic card for about $40,[35] now restricted to compatible devices and without the ability to copy the files.

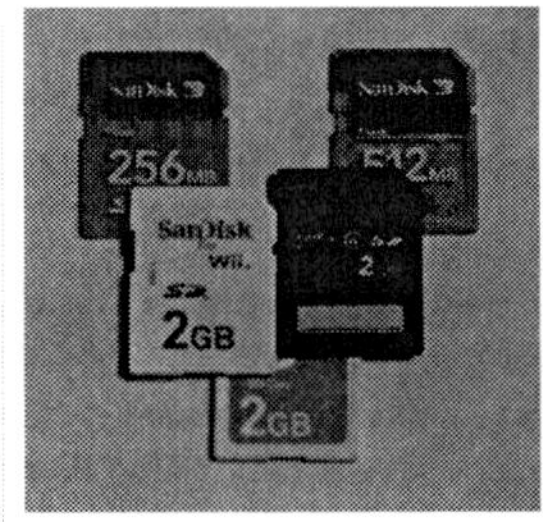

SanDisk cards in various colors

- **Integrated USB connector** — The SanDisk **SD Plus** product can be plugged directly into a USB port without needing a USB card reader.[36] Other companies introduced comparable products, such as the **Duo SD** product of OCZ Technology and the **3 Way** (microSDHC, SDHC, and USB) product of A-DATA, which was available in 2008 only.
- **Different colors** — SanDisk has used various colors of plastic or adhesive label, including a "gaming" line in translucent plastic colors that indicated the card's capacity.
- **Integrated display** — In 2006, A-DATA announced a **Super Info SD** card with a digital display that provided a two-character label and showed the amount of unused memory on the card.[37]

Speeds

An SD card's speed is measured by how quickly information can be read from, or written to, the card. In applications that require sustained write throughput, such as video recording, the device might not perform satisfactorily if the SD card's class rating falls below a particular speed. For example, a camcorder built for a Class 6 card may suffer dropouts or corrupted video if a slower card is used. Digital cameras may experience a noticeable lag between shots, while the camera writes the picture to a slower card.

A card's speed depends on many factors, such as the following:

- The likelihood of soft errors that the card's controller must re-try
- The fact that, on most cards, writing data requires the controller to read and erase a larger region, then rewrite that entire region with the desired part changed
- The possibility of fragmentation: that a body of information the host views as a unit is, for historical reasons, written to non-contiguous regions of memory. (This possibility does not cause rotational or head-movement delays as with magnetic media, but it does vary the amount of computation the card's controller must do.)

In early SD cards, the speed was measured with the × rating, which compared the average speed of reading data to that of the original CD-ROM drive. Currently, the official unit of measurement is the **Speed Class Rating**, which guarantees a minimum rate at which data can be written to the card.

The newer families of SD card improve card speed by increasing the bus rate (the frequency of the clock signal that strobes information into and out of the card). Whatever the bus rate, the card can signal to the host that it is "busy" until a read or a write operation is complete. Compliance with a higher speed rating is a guarantee that the card limits its use of the "busy" indication.

Speed Class Rating

32GB SDHC card

The **Speed Class Rating** is the official unit of speed measurement for SD cards. The class number guarantees a minimum write speed as a multiple of 8 Mbit/s (1 MB/s). The SDA defines several speed class ratings, but manufacturers may claim conformance to those ratings without independent verification.

The host device can read a card's speed class, unlike the earlier "×" speed ratings. A device can warn the user if the card reports a speed class that falls below an application's minimum need.[38]

These are the ratings of all currently available cards:[39] [40]

Class	Speed
Class 2	2 MB/s
Class 4	4 MB/s
Class 6	6 MB/s
Class 10	10 MB/s

Speed Classes 2, 4, and 6 assert that the card supports the respective number of MB/s as a minimum sustained write speed for a card in a fragmented state. Class 10 asserts that the card supports 10 MB/s as a minimum non-fragmented sequential write speed.[38] By comparison, the older "×" rating measured maximum speed under ideal conditions, and was vague as to whether this was read speed or write speed.

× rating

The × rating is a multiple of the standard CD-ROM drive speed of 1.2 Mbit/s (approximately 150 kB/s). Basic cards transfer data up to six times (6×) the CD-ROM speed; that is, 7.2 Mbit/s. The 2.0 specification defines speeds up to 200×, but is not as specific as Speed Classes are on how to measure speed. Manufacturers may report best-case speeds and may report the card's fastest read speed, which is typically faster than the write speed. Vendors including Transcend and Kingston report their cards' write speed.[41]

This table lists common ratings, the minimum transfer rates, and the corresponding Speed Class (though the comparison is not always exact).

Rating	Read Speed (MB/s)	Write Speed (MB/s)	Speed Class
6×	0.9		
10×	1.5		
13×	**2.0**	**2.0**	**2**
26×	**4.0**	**4.0**	**4**
32×	4.8	5.0	
40×	**6.0**	**6.0**	**6**
66×	**10.0**	**10.0**	**10**
100×	15.0	15.0	
133×	20.0	20.0	
150×	22.5	22.5	
200×	30.0	30.0	
266×	40.0	40.0	
300×	45.0	45.0	
400×	60.0	60.0	
600×	90.0	90.0	

UHS Speed Class

The Ultra-High Speed (UHS) interface is available on some SDHC and SDXC cards.[42] The following ultra-high speeds are specified:

- **UHS-I** cards, specified in SD Version 3.01,[38] support a clock frequency of 100 MHz (a quadrupling of the original Default Speed), which in four-bit transfer mode could transfer 50 MB/s. UHS-I cards declared as **UHS104** also support a clock frequency of 208 MHz, which could transfer 104 MB/s. UHS-I is the only class for which products are currently available.[40]
- Double data rate operation at 50 MHz (DDR50) is also specified in Version 3.01, and is mandatory for microSDHC and microSDXC cards labeled as UHS-I. In this mode, four bits are transferred when the clock signal rises and another four bits when it falls, transferring an entire byte on each full clock cycle.
- **UHS-II** cards, to be defined in Version 4.0, further raise the data transfer rate to a theoretical maximum of 312 MB/s.[26] [43]

UHS memory cards work best with UHS host devices. The combination lets the user record HD resolution videos to tapeless camcorders while performing other functions. It is also suitable for real-time broadcasts and capturing large HD videos.

Cards that comply with UHS show UHS-I or UHS-II on the label, and report this capability to the host device. Use of UHS requires that the host device command the card to drop from 3.3-volt to 1.8-volt operation and select the 4-bit transfer mode.

Market penetration

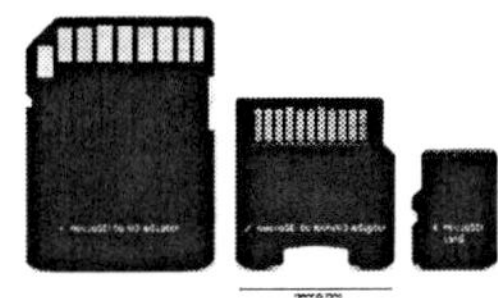
microSD to SD adapter (left), microSD to miniSD adapter (middle), microSD card (right)

Secure Digital cards are used in many consumer electronic devices, and have become a widespread means of storing several gigabytes of data in a small size. Devices where the user may remove and replace cards often, such as digital cameras, camcorders, and video game consoles, tend to use full-sized cards. Devices where small size is paramount, such as mobile phones, tend to use microSD cards. SD cards are not the most economical solution in devices that need only a small amount of non-volatile memory, such as station presets in small radios. They may also not present the best choice for applications where higher storage capacities or speeds are a requirement as provided by other flash card standards such as Compact Flash.

Many personal computers of all types and personal digital assistants (PDAs) use SD cards, either through built-in slots or through an active electronic adaptor. Adaptors exist for the PC card, ExpressBus, USB, FireWire, and the parallel printer port. Active adaptors also let SD cards be used in devices designed for other formats. such as CompactFlash. The FlashPath adaptor lets SD cards be used in a floppy disk drive.

Digital cameras

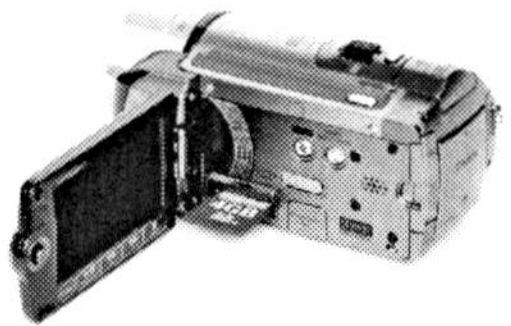
A camcorder with a 4 GB SDHC card

SD/MMC cards replaced Toshiba's SmartMedia as the dominant memory card format used in digital cameras. In 2001, SmartMedia had achieved nearly 50% use, but by 2005 SD/MMC had achieved over 40% of the digital camera market and SmartMedia's share had plummeted, with cards not being easily available in 2007.

At this time all the leading digital camera manufacturers use SD in their consumer product lines, including Canon, Casio, Fujifilm, Kodak, Leica, Nikon, Olympus, Panasonic, Pentax, Ricoh, Samsung, and Sony. Formerly, Olympus and Fujifilm used XD-Picture Cards (xD cards) exclusively, while Sony only used Memory Stick; however as of January 2010, all three support SD.

Some prosumer and professional digital camera models continue to offer CompactFlash, either on a second card slot or as the only storage, as they offer much higher capacities and faster transfer speeds and historically offered a better price/capacity ratio as well.

Secure Digital memory cards can be used in Sony XDCAM EX camcorders via the MEAD-SD01 adapter.[44]

Personal computers

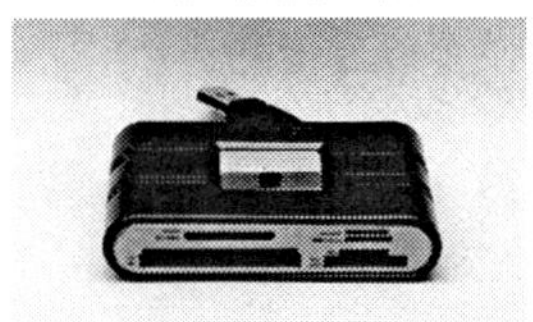
USB-based universal Memory card reader

Although many personal computers accommodate SD cards as an auxiliary storage device through a built-in slot or a USB adaptor, SD cards cannot be used as the primary hard disk through the onboard ATA controller because none of the SD card variants supports ATA signalling. This use requires a separate SD controller chip[45] or a SD-to-CompactFlash converter. However, on computers that support bootstrapping from a USB interface, an SD card in a USB adaptor can be the primary hard disk, provided it contains an operating system that supports USB access once the bootstrap is complete.

Embedded systems

In 2008, the SDA specified Embedded SD, "leverag[ing] well-known SD standards" to enable non-removable SD-style devices on printed circuit boards.[46] SanDisk provides such memory components under the iNAND brand.[47]

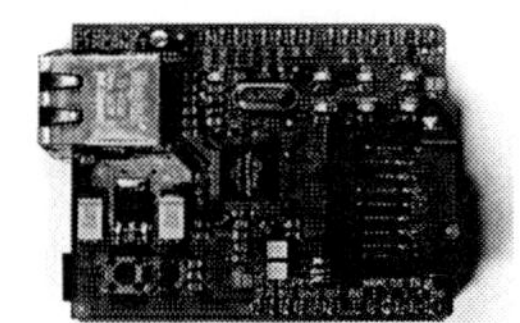

This shield (daughterboard) gives Arduino prototyping microprocessors access to SD cards plugged into the socket.

Most modern microcontrollers have built-in SPI logic that can interface to a SD card operating in its SPI mode, providing non-volatile storage. Even if a microcontroller lacks the SPI feature, the feature can be emulated by bit banging. For example, a home-brew hack combines spare General Purpose Input/Output (GPIO) pins of the processor of the Linksys WRT54G router with MMC support code from the Linux kernel.[48] This technique can achieve throughput of up to 1.6 Mbit/s.

History

In 1999, SanDisk, Matsushita, and Toshiba agreed to develop and market the Secure Digital (SD) Memory Card, which was a development of the MultiMediaCard (MMC). The new card provided both digital rights management (DRM) up to the Secure Digital Music Initiative (SDMI) standard, and a high memory density for the time.

On top: a microSDHC card that stores about 8 billion bytes. Below it: 64 units of magnetic-core memory, used until the 1970s, that stored eight bytes.

The new format was designed to compete with the Memory Stick, a DRM product that Sony released the prior year. It was mistakenly predicted that DRM features[49] would be widely used due to pressure from music and other media suppliers to prevent piracy.

The trademarked *SD* logo was originally developed for the Super Density Disc, which was the unsuccessful Toshiba entry in the DVD format war. This is why the *D* resembles an optical disc.

At the 2000 Consumer Electronics Show (CES) trade show, the three companies announced the creation of the SD Card Association (SDA) to promote SD cards. The association's headquarters are in California and it comprises some 30 companies that produce devices and content. Early samples of the SD Card were available in the first quarter of 2000, with production quantities of 32 and 64 MB cards available 3 months later.

Relation to MMC

The SD cards changed the MMC design in several ways:

- Asymmetrical slots in the sides of the SD card prevent inserting it upside down, while an MMC will go in most of the way but make no contact if inverted.
- Most SD cards are 2.1 mm thick, compared to 1.4 mm for MMCs. The SD specification defines a card called **Thin SD** with a thickness of 1.4 mm, but they are rare, as the SDA went on to define even smaller form factors.
- The card's electrical contacts are recessed beneath the surface of the card, protecting them from contact with a user's fingers.
- The SD specification envisaged capacities and transfer rates exceeding those of MMC, and these have both grown over time. For a comparison table, see below.

Size comparison of various flash cards: SD, CompactFlash, MMC, xD

Full-sized SD cards will not fit into the slimmer MMC slots, and there are other issues that affect the ability to use one format in a host device designed for the other.

Follow-on products

In March 2003, SanDisk Corporation announced the introduction of the **miniSD** and demonstrated it at CeBIT 2003.[50] The SDA adopted the miniSD card in 2003 as a small form factor extension to the SD card standard. While the new cards were designed especially for use in mobile phones, they are usually packaged with a miniSD adapter which enables compatibility with all devices equipped with a standard SD memory card slot.

At CTIA Wireless 2005, the SDA announced the small **microSD** form factor (and SDHC, with capacities in excess of 2 GB and a minimum sustained read and write speed of 17.6 Mbit/s). SanDisk had conceived microSD when its CTO and the CTO of Motorola concluded that current memory cards were too large for mobile phones. The card was originally called T-Flash, but just before product launch, T-Mobile sent a cease and desist order to SanDisk claiming that T-Mobile owned the trademark on T-(anything), and the name was changed to TransFlash. TransFlash and microSD cards are the same; each can be used in devices made for the other.[51] SanDisk induced the SDA to administer the microSD standard. The SDA approved the final microSD specification on July 13, 2005. Initially, microSD cards were available in capacities of 32, 64, and 128 MB.

In April 2006, the SDA released a detailed specification for the non-security related parts of the SD memory card standard and for the Secure Digital Input Output (SDIO) cards and the standard SD host controller.

In September, 2006, SanDisk announced the 4 GB miniSDHC.[52] Like the SD and SDHC, the miniSDHC card has the same form factor as the older miniSD card but the HC card requires HC support built into the host device. Devices that support miniSDHC will work with miniSD and miniSDHC, but devices without specific support for miniSDHC will work only with the older miniSD card.

In January 2009, the SDA announced the SDXC family, which supports cards up to 2 TB and speeds up to 300 Mbyte/s.

Openness of the standard

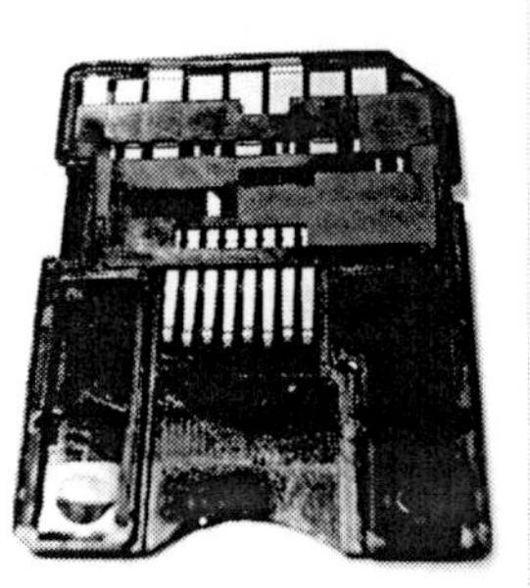

Dismantled microSD to SD adapter showing the passive connection from the microSD card slot on the bottom to the SD pins on the top

Like most memory card formats, SD is covered by numerous patents and trademarks. Royalties for SD card licences are imposed for manufacture and sale of memory cards and host adapters (1000 USD/year plus membership at 1500 USD/year), but SDIO cards can be made without royalties.

Early versions of the SD specification were available only after agreeing to a non-disclosure agreement (NDA) that prohibited development of an open source driver. However, the system was eventually reverse-engineered, and free software drivers provided access to SD cards that did not use DRM. Since then, the (SDA) has provided a simplified version of the specification under a less restrictive license.[53] Although most open-source drivers were written before this, it has helped them solve compatibility issues.

In 2006, the SDA also released a simplified version of the specification of the host controller interface (as opposed to the specification of SD cards) and later also for the physical layer, ASSD extensions, SDIO, and SDIO Bluetooth Type-A, under a disclaimers agreement.[54] Again, most of the information had already been discovered and Linux had a fully free driver for it. Still, building a chip conforming to this specification caused the One Laptop per Child project to claim "the first truly Open Source SD implementation, with no need to obtain an SDI license or sign NDAs to create SD drivers or applications."[55]

The fact that the complete SD specification is proprietary mainly affects embedded systems and laptops, since users of desktop PCs generally read SD cards via USB-based card readers. These card readers present a standard USB mass storage interface to memory cards, thus separating the operating system from the details of the underlying SD interface. However, embedded systems (such as portable music players) usually gain direct access to SD cards and thus need complete programming information. Desktop card readers are themselves embedded systems; their manufacturers have usually paid the SDA for complete access to the SD specifications. Many notebook computers now include SD card readers not based on USB; device drivers for these essentially gain direct access to the SD card, as do embedded systems.

Technical details

Transfer modes

The physical interface comprises 9 pins, except that the miniSD card adds two unconnected pins in the center and the microSD card omits one of the two V_{SS} (Ground) pins.

SD card pin assignment	miniSD card pin assignment	microSD card pin assignment

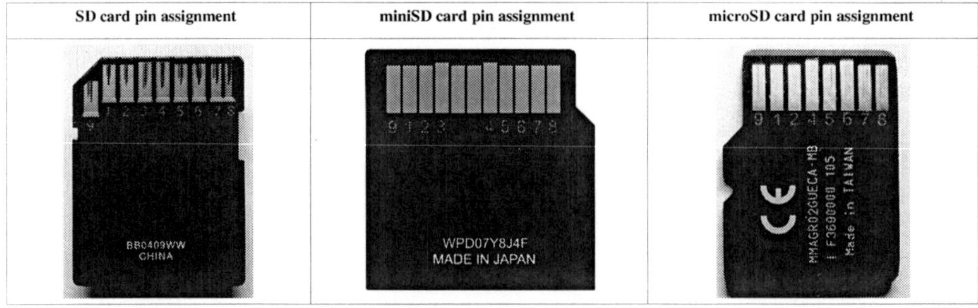

Various SD cards may support various combinations of the following bus types and transfer modes. The SPI bus and one-bit SD bus are mandatory for all SD families, as explained in the next section.

- **SPI:** Serial Peripheral Interface Bus is primarily used by embedded microcontrollers. This bus type supports only a 3.3-volt interface.
- **One-bit SD:** Separate command and data channels and a proprietary transfer format.
- **Four-bit SD:** Uses extra pins plus some reassigned pins. UHS-I and UHS-II requires this bus type.

Once the host device and the SD card negotiate a bus interface, the usage of the numbered pins is the same for all three card sizes:

SPI Bus

Pin	Name	I/O	Logic	Description
1	nCS	I	PP	Card Select (Neg True)
2	DI	I	PP	Data In [MOSI]
3	VSS	S	S	Ground
4	VDD	S	S	Power
5	CLK	I	PP	Clock [SCLK]
6	VSS	S	S	Ground
7	DO	O	PP	Data Out [MISO]
8	NC nIRQ	. O	. OD	NC (Memory Cards) Interrupt (SDIO Cards)
9	NC	.	.	NC

One-Bit SD Bus

Pin	Name	I/O	Logic	Description
1	NC	.	.	NC
2	CMD	I/O	PP,OD	Command, Response
3	VSS	S	S	Ground
4	VDD	S	S	Power
5	CLK	I	PP	Clock
6	VSS	S	S	Ground
7	DAT0	I/O	PP	Data 0
8	NC nIRQ	. O	. OD	NC (Memory Cards) Interrupt (SDIO Cards)
9	NC	.	.	NC

Four-Bit SD Bus

Pin	Name	I/O	Logic	Description
1	DAT3	I/O	PP	Data 3
2	CMD	I/O	PP,OD	Command, Response
3	VSS	S	S	Ground
4	VDD	S	S	Power
5	CLK	I	PP	Clock
6	VSS	S	S	Ground
7	DAT0	I/O	PP	Data 0
8	DAT1 nIRQ	I/O O	PP OD	Data 1. SDIO Cards share with Interrupt Period
9	DAT2	I/O	PP	Data 2

Notes:

1. Direction is relative to card. I = Input, O = Output.
2. PP = Push-Pull logic, OD = Open-Drain logic.
3. S = Power Supply, NC = Not Connected (or logical high).

Interface

Command interface

Inside a 512 MB SD card: NAND flash chip that holds the data (bottom) and SD controller (top)

SD cards and host devices initially communicate through a synchronous one-bit interface, where the host device provides a clock signal that strobes single bits into and out of the SD card. The host device thereby sends 48-bit commands and receives responses. The card can signal that a response will be delayed, but the host device can abort the dialogue.

Through issuing various commands, the host device can:

- Determine the type, memory capacity, and capabilities of the SD card
- Command the card to use a different voltage, different clock speed, or advanced electrical interface
- Prepare the card to receive a block to write to the flash memory, or read and reply with the contents of a specified block.

The command interface is an extension of the MultiMediaCard (MMC) interface. SD cards dropped support for some of the commands in the MMC protocol, but added commands related to copy protection. By using only commands supported by both standards until determining the type of card inserted, a host device can accommodate both SD and MMC cards.

Electrical interface

All SD card families initially use a 3.3-volt electrical interface. On command, SDHC and SDXC cards switch to 1.8-volt operation.[38]

At initial power-up or card insertion, the host device selects either the Serial Peripheral Interface (SPI) bus or the one-bit SD bus by the voltage level present on Pin 1. Thereafter, the host device may issue a command to switch to the four-bit SD bus interface, if the SD card supports it. For various card types, support for the four-bit SD bus is either optional or mandatory.[38]

After determining that the SD card supports it, the host device can also command the SD card to switch to a higher transfer speed. Until determining the card's capabilities, the host device should not use a clock speed faster than 400 kHz. SD cards other than SDIO (see below) have a Default Speed clock rate of 25 MHz. The host device is not required to use the maximum clock speed that the card supports. It may operate at less than the maximum clock speed to conserve power.[38] Between commands, the host device can stop the clock entirely.

Inside a 2 GB SD card: two NAND flash chips (top and middle), SD controller chip (bottom)

SDIO cards

The SDIO family comprises Low-Speed and Full-Speed cards. Both types of SDIO cards support SPI and one-bit SD bus types. Low-Speed SDIO cards are allowed to also support the four-bit SD bus; Full-Speed SDIO cards are required to support the four-bit SD bus. To use a SDIO card as a "combo card" (for both memory and I/O), the host device must first select four-bit SD bus operation. Two other unique features of Low-Speed SDIO are a maximum clock rate of 400 kHz for all communications, and the use of Pin 8 as "interrupt" to try to initiate dialogue with the host device.[56]

Ganging cards together

Inside a SDHC card

The one-bit SD protocol was derived from the MMC protocol, which envisaged the ability to put up to 3 cards on a bus of common signal lines. The cards use open collector interfaces, where a card may pull a line to the low voltage level; the line is at the high voltage level (because of a pull-up resistor) if no card pulls it low. Though the cards shared clock and signal lines, each card had its own chip select line to sense that the host device had selected it.

The SD protocol envisaged the ability to gang 30 cards together without separate chip select lines. The host device would broadcast commands to all cards and identify the card to respond to the command using its unique serial number.

In practice, cards are rarely ganged together because open-collector operation has problems at high speeds and increases power consumption. Newer versions of the SD specification recommend separate lines to each card.

Achieving higher card speeds

The SD specification defines four-bit-wide transfers. (The MMC specification supports this and also defines an eight-bit-wide mode.) Transferring several bits on each clock pulse improves the card speed. Advanced SD families have also improved speed by offering faster clock frequencies and double data rate (explained here).

File system

Like other types of flash memory card, an SD card of any SD family is a block-addressable storage device, in which the host device can read or write fixed-size blocks by specifying their block number.

MBR and FAT

Most SD cards ship preformatted with one or more MBR partitions, where the first or only partition contains a file system. This lets them operate like the hard disk of a personal computer. Per the SD card specification, an SD card is formatted with MBR and the following file system:

- For SDSC cards: FAT16
- For SDHC cards: FAT32
- For SDXC cards: exFAT

Most consumer products that take an SD card will expect it to be partitioned and formatted in this way. The universal support for FAT16 and FAT32 allow the usage of SDSC and SDHC cards on most host computers with a compatible SD reader, to present the user with the familiar method of named files in a hierarchical directory tree.

On such SD cards, standard utility programs such as Mac OS X's "Disk Utility" or Windows' SCANDISK can be used to repair or retrieve corrupted data, and sometimes recover deleted files. Defragmentation tools for FAT file systems may be used on such cards. The resulting consolidation of files may provide a marginal improvement in the time required to read or write the file,[39] but not an improvement comparable to defragmentation of hard drives, where storing a file in multiple fragments may involve a time penalty to move between physical areas of the drive. Moreover, defragmentation performs writes to the SD card that count against the card's rated lifespan. The write endurance of the physical memory is discussed in the article on flash memory; newer technology to increase the storage capacity of a card currently provides worse write endurance.

When reformatting an SD card smaller than 4 GB, FAT16 should be used. (This is also an option for 4 GB cards, but it requires the use of 64 kiB clusters, which are not widely supported.) FAT16 does not support cards above 4 GB.

The SDXC specification makes Microsoft's proprietary exFAT file system mandatory,[57] which is supported only by some proprietary operating systems.

Other file systems

Because the host views the SD card as a block storage device, the card does not require MBR partitions or any specific file system. The card can be reformatted to use any file system the operating system supports. For example:

- Under Unix-like operating systems such as Linux or FreeBSD, SD cards can be formatted using the UFS, EXT3, ext4, btrFS, HFS Plus, or the ReiserFS file systems.
- Under Mac OS X, SD cards can be partitioned as GUID devices and formatted with the HFS Plus file system.
- Under Windows and some Unix systems, SD cards can be formatted using NTFS and, on later versions, exFAT.

Additionally, an SD card called Live SD can contain an embedded operating system (such as Live USB). Computers that can bootstrap from an SD card (either using a USB adapter or inserted into the computer's flash media reader) instead of the hard disk drive may thereby be able to recover from a corrupted hard disk drive. A Live SD can be write-locked to preserve the system's integrity.

Risks of reformatting

Reformatting an SD card with a different file system, or even with the same one, may make the card slower, or shorten its lifespan. Some cards use wear leveling, in which frequently modified blocks are mapped to different portions of memory at different times, and some wear-leveling algorithms are designed for the access patterns typical of the file allocation table on a FAT16 or FAT32 device.[58] In addition, the preformatted file system may use a cluster size that matches the erase region of the physical memory on the card; reformatting may change the cluster size and make writes less efficient.

Power use

The power consumption of microSD cards varies by manufacturer, but appears to be in the range of 66-330 mW (20-100 mA at a supply voltage of 3.3 V). Specifications from TwinMos technologies list a maximum of 149 mW (45 mA) during transfer. Toshiba, on the other hand, lists 264-330 mW (80-100 mA).[59]

Storage capacity and incompatibilities

All SD cards let the host device determine how much information the card can hold, and the specification of each SD family gives the host device a guarantee of the maximum capacity a compliant card will report.

By the time the Version 2.0 (SDHC) specification was completed in June 2006,[60] vendors had already devised 2 GB and 4 GB SD cards, either as specified in Version 1.01, or by creatively reading Version 1.00. The resulting cards do not work correctly in some host devices.[61] [62]

SDSC cards above 1 GB

A host device can ask any inserted SD card for its 128-bit identification string (the Card-Specific Data or CSD). In standard-capacity cards (SDSC), 12 bits identify the number of memory clusters (ranging from 1 to 4,096) and 3 bits identify the number of blocks per cluster (which decode to 4, 8, 16, 32, 64, 128, 256, or 512 blocks per cluster). The host device multiplies these figures (as shown in the following section) with the number of bytes per block to determine the card's capacity in bytes.

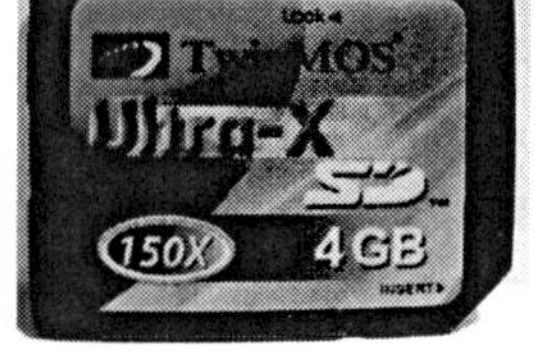

4 GB SDSC card

In SD version 1.00, the number of bytes per block was assumed to be 512. This permitted SDSC cards up to 4,096 × 512 × 512 = 1 GB, for which there are no known incompatibilities.

Version 1.01 let an SDSC card use a 4-bit field to indicate 1,024 or 2,048 bytes per block instead.[38] Doing so enabled cards with 2 GB and 4 GB capacity.

Early SDSC host devices that assume 512-byte blocks therefore will not fully support the insertion of 2 GB or 4 GB cards. In some cases, the host device can read data that happens to reside in the first 1 GB of the card. If the assumption is made in the driver software, success may differ from one version of Windows to another. In addition, any host device might not support a 4 GB SDSC card, since the specification lets it assume that 2 GB is the maximum for these cards.

Storage capacity calculations

The format of the Card-Specific Data (CSD) register changed between version 1 (SDSC) and version 2.0 (which defines SDHC and SDXC).

Version 1

In Version 1 of the SD specification, capacity is calculated by combining fields of the CSD as follows:

```
Capacity=(C_SIZE+1)<<(C_SIZE_MULT+2)<<READ_BL_LEN        2 GiB max.
Where 0<=C_SIZE<=4095, 0<=C_SIZE_MULT<=7, READ_BL_LEN==9 || READ_BL_LEN==10
```

Later versions state (at Section 4.3.2) that a 2 GB SDSC card shall set its READ_BL_LEN (and WRITE_BL_LEN) to indicate 1024 bytes, so that the above computation correctly reports the card's capacity; but that, for consistency, the host device shall not request (by CMD16) block lengths over 512 bytes.[38]

Versions 2 and 3

In the definition of SDHC cards in Version 2.0, the C_SIZE portion of the CSD is 22 bits and it indicates the memory size in multiples of 512 KB. (The C_SIZE_MULT field is removed and READ_BL_LEN is no longer used to compute capacity.) Two bits that were formerly reserved now identify the card family: 0 is SDSC; 1 is SDHC or SDXC; 2 and 3 are reserved.[38] Because of these redefinitions, older host devices do not correctly identify SDHC or SDXC cards nor their correct capacity.

- SDHC cards are restricted to reporting a capacity not over 32 GiB.
- SDXC cards are allowed to use all 22 bits of the C_SIZE field. An SDHC card that did so (reported C_SIZE > 65375 to indicate a capacity of over 32 GiB) would violate the specification. A host device that relied on C_SIZE rather than the specification to determine the card's maximum capacity might support such a card, but the card might fail in other SDHC-compatible host devices.

Capacity is calculated thus:

```
Capacity=(C_SIZE+1)*524288
where for SDHC        4112<=C_SIZE<=65375      (approx. 2 GB) < capacity < 32 GiB
      for SDXC       65535<=C_SIZE                       32 GiB <= capacity <= 2 TiB max.
```

Capacities above 4 GB can only be achieved by following Version 2.0 or later versions. In addition, capacities equal to 4 GB must also do so to guarantee compatibility.

Comparison to other flash memory formats

Overall, SD is less open than CompactFlash or USB flash memory drives; these are open standards which can be implemented free of payment for licensing, royalties, or documentation. (CompactFlash and USB flash drives may, however, require licensing fees for the use of the SDA's trademarked logos.)

However, SD is much more open than Memory Stick, for which no public documentation nor any documented legacy implementation is available. All SD cards can be accessed freely using the well-documented SPI bus.

xD cards are simply 18-pin NAND flash chips in a special package and support the standard command set for raw NAND flash access. Although the raw hardware interface to xD cards is well understood, the layout of its memory contents—necessary for interoperability with xD card readers and digital cameras—is totally undocumented. The consortium that licenses xD cards has not released any technical information to the public.

Type	MMC	RS-MMC	MMC Plus	SecureMMC	SD	SDIO	miniSD	microSD
SD Socket	Yes	Mechanical adapter	Yes	Yes	Yes	Yes	Electromechanical adapter	Electromechanical adapter
Pins	7	7	13	7	9	9	11	8
Form factor	shallow	shallow/narrow	shallow	shallow	deep (some)	deep	narrow/slim/shallow	narrow/slim/extra shallow
Breadth	24 mm	24 mm	24 mm	24 mm	24 mm	24 mm	20 mm	11 mm
Width	32 mm	18 mm	32 mm	32 mm	32 mm	32 mm+	21.5 mm	15 mm
Depth	1.4 mm	1.4 mm	1.4 mm	1.4 mm	2.1 mm (some)	2.1 mm	1.4 mm	1 mm
SPI mode	Optional	Optional	Optional	Yes	Yes	Yes	Yes	Yes
1-bit mode	Yes	Yes	Yes	Yes	Yes	Yes	Yes	Yes
4-bit mode	No	No	Yes	?	Optional	Optional	Optional	Optional
8-bit mode	No	No	Yes	No	No	No	No	No

Interrupts	No	No	No	No	No	Optional	No	No
Max clock rate	20 MHz	20 MHz	52 MHz	20 MHz?	208 MHz	50 MHz	208 MHz	208 MHz
Max transfer	20 Mbit/s	20 Mbit/s	416 Mbit/s	20 Mbit/s?	832 Mbit/s	200 Mbit/s	832 Mbit/s	832 Mbit/s
Max SPI transfer	20 Mbit/s	20 Mbit/s	52 Mbit/s	20 Mbit/s	50 Mbit/s	50 Mbit/s	50 Mbit/s	50 Mbit/s
DRM	No	No	No	Yes	Yes	N/A	Yes	Yes
User encrypt	No	No	No	Yes	No	No	No	No
Simplified spec	Yes	Yes	No	Not yet?	Yes	Yes	No	No
Membership cost	JEDEC: $4400/yr, optional				SD Card Association: $2000/yr, general; $4500/yr, executive			
Specification cost	Free			?	Simplified spec: free Full spec: free to members, $1000/yr to R&D non-members			
Host license	No	No	No	No	Yes $1000/yr			
Card royalties	Yes	Yes	Yes	Yes	Yes	Yes +$1000/yr	Yes	Yes
Open source compatible	Yes	Yes	Yes?	Yes?	Yes	Yes	Yes	Yes
Nominal operating voltage	3.3V	1.8V/3.3V	1.8V/3.3V[63][64]	1.8V/3.3V	3.3V	3.3V	3.3V	3.3V
Type	**MMC**	**RS-MMC**	**MMC Plus**	**SecureMMC**	**SD**	**SDIO**	**miniSD**	**microSD**

Table data compiled mostly from simplified versions of MMC and SDIO specifications and other data on SD card and MMC association web sites. Data for other card variations is interpolated.

Capacity limit in all SD/MMC formats appears to be 128 GB in LBA mode (28-bit sector address).

See also

- SD Card Association
- Comparison of memory cards
- Serial Peripheral Interface Bus (SPI)
- File Allocation Table (FAT16, FAT32) and exFAT
- Flash memory
- slotMusic
- MultiMediaCard
- Universal Flash Storage

References

[1] "About" (http://www.sdcard.org/developers/about/). SD Card Association. . Retrieved 2011-12-08.

[2] "Capacity" (http://www.sdcard.org/developers/overview/capacity/). SD Card Association. . Retrieved 2011-12-08.

[3] "Using SDXC" (http://www.sdcard.org/consumers/sdxc_capabilities/using_sdxc/). SD Card Association. . Retrieved 2011-12-08.

[4] "SDIO" (http://www.sdcard.org/developers/overview/sdio/). SD Card Association. . Retrieved 2011-12-08.

[5] "Using SD Memory Cards is Easy" (video on YouTube) - SD Card Association (http://www.youtube.com/watch?v=wX6NwBa1csY)

[6] What are SDHC, miniSDHC, and microSDHC? (http://www.sandisk.com/Assets/File/pdf/retail/SDHC1.pdf) SanDisk.com

[7] What's new in Firmware 2.41 Beta (for COWON D2) (http://www.jetaudio.com/download/cowon_rn_d2.html#241) JetAudio

[8] "Microsoft Support 934428 - Hotfix for Windows XP that adds support for SDHC cards that have a capacity of more than 4 GB" (http://support.microsoft.com/kb/934428). Support.microsoft.com. 2008-02-15. . Retrieved 2010-08-22.

[9] "Microsoft Support 939772 - Some Secure Digital (SD) cards may not be recognized in Windows Vista" (http://support.microsoft.com/kb/939772). Support.microsoft.com. 2008-05-15. . Retrieved 2010-08-22.

[10] "Microsoft Support 949126 - A Secure Digital High Capacity (SDHC) card is not recognized on a Windows Vista Service Pack 1-based computer" (http://support.microsoft.com/kb/949126). Support.microsoft.com. 2008-02-21. . Retrieved 2010-08-22.

[11] "SanDisk and Sony to expand Memory Stick Pro and Memory Stick Micro formats" (http://www.sandisk.com/about-sandisk/press-room/press-releases/2009/2009-01-07-sandisk-and-sony-to-expand-âmemory-stick-proâ-and-âmemory-stick-microâ-formats). Sandisk.com. . Retrieved 2010-08-22.

[12] "SD Card, Memory Stick formats to reach 2 terabytes, but when? - Betanews" (http://www.betanews.com/article/SD_Card_Memory_Stick_formats_to_reach_2_terabytes_but_when/1231453659). . 090108 betanews.com

[13] "Pretec introduces world's first SDXC card: Digital Photography Review" (http://www.dpreview.com/news/0903/09030601pretecsdxc.asp). Dpreview.com. 2009-03-06. . Retrieved 2010-08-22.

[14] "Canon EOS Rebel T2i / 550D Digital SLR Camera Review" (http://www.the-digital-picture.com/Press-Release/Canon-EOS-Rebel-T2i-550D-Digital-SLR-Camera-Press-Release.aspx). . The-Digital-Picture.com

[15] Ng, Jansen (2009-11-24). "Lack of Card Readers Holding Back SDXC Flash Memory Adoption" (http://www.dailytech.com/Lack+of+Card+Readers+Holding+Back+SDXC+Flash+Memory+Adoption/article16915.htm). DailyTech. . Retrieved 2009-12-22.

[16] Ng, Jansen (2009-11-30). "Lenovo, HP, Dell Integrating SDXC Readers in New 32nm Intel "Arrandale" Laptops" (http://www.dailytech.com/Lenovo+HP+Dell+Integrating+SDXC+Readers+in+New+32nm+Intel+Arrandale+Laptops/article16937.htm). DailyTech. . Retrieved 2009-12-22.

[17] Ng, Jansen (2009-12-22). "Toshiba Sampling First SDXC Flash Memory Cards" (http://www.dailytech.com/Toshiba+Sampling+First+SDXC+Flash+Memory+Cards/article16972.htm). DailyTech. . Retrieved 2009-12-22.

[18] "Toshiba's 64 GB SDXC card to finally go on sale (in Japan)" (http://www.crunchgear.com/2010/04/15/toshibas-64gb-sdxc-card-to-finally-go-on-sale-in-japan/). CrunchGear. . Retrieved 2010-08-09.

[19] "Panasonic Introduces New 64 GB* and 48 GB* SDXC Memory Cards, Available Globally in February 2010" (http://www2.panasonic.com/webapp/wcs/stores/servlet/prModelDetail?storeId=11301&catalogId=13251&itemId=389511&modelNo=Content01052010041118461&surfModel=Content01052010041118461). Panasonic. . Retrieved 2010-08-09.

[20] "Sandisk ships its highest capacity sd card ever" (http://sandisk.com/about-sandisk/press-room/press-releases/2010/2010-02-22-sandisk-ships-its-highest-capacity-sd-card-ever,-the-64gb-sandisk-ultra-sdxc-card). SanDisk. . Retrieved 2010-08-09.

[21] Lexar ships 128 GB Class 10 SDXC card; March 2011. (http://www.betanews.com/article/Lexar-ships-first-128GB-SDXC-cards/1300305310)

[22] "SDXC/SDHC 433X Class 16 Card from Pretec" (http://www.pretec.com/news-event/press-room/item/root/sdxcsdhc-433x-c16). Pretec. 2011-06-13. . Retrieved 2010-12-03.

[23] First 64GB microSD Card Here; When Will Smartphones Support It? (http://pocketnow.com/smartphone-news/first-64gb-microsd-card-arrives-when-will-smartphones-support)

[24] Kingmax flaunts world's first 64GB microSD card (http://www.engadget.com/2011/05/26/kingmax-flaunts-worlds-first-64gb-microsd-card/)

[25] About Compatibility with Host Devices (https://www.sdcard.org/consumers/compatibility/) SD Association

[26] "SD Card Association Triples Speeds with UHS-II" (https://www.sdcard.org/press/SD_Association_Announces_UHS-II_eBOOK_Jan_5_2011_ENGLISH.PDF). . Retrieved 2011-08-09.

[27] http://www.microsoft.com/downloads/details.aspx?FamilyID=1cbe3906-ddd1-4ca2-b727-c2dff5e30f61&displaylang=en

[28] "About the SD and SDXC card slots" (http://support.apple.com/kb/HT3553). Apple Inc.. 2011-05-03. . Retrieved 2011-09-05.

[29] "Apple released exFAT support in OS X 10.6.5 update" (http://www.tuxera.com/mac/apple-released-exfat-support-in-os-x-10-6-5-update/). Tuxera.com. 2010-11-22. . Retrieved 2012-01-04.

[30] "Windows Phone 7 Secure Digital Card Limitations" (http://support.microsoft.com/kb/2450831). .

[31] "Windows Phone 7's microSD mess: the full story (and how Nokia can help you out of it)" (http://www.engadget.com/2010/11/17/windows-phone-7s-microsd-mess-the-full-story-and-how-nokia-ca). .

[32] "Super Talent Technology - DDR and DDR2 Memory" (http://www.supertalent.com/products/sd.php). Supertalent.com. . Retrieved 2010-08-22.

[33] "Home" (http://www.eye.fi). Eye-Fi. . Retrieved 2010-08-22.

[34] AudioHolics (http://www.audioholics.com/news/industry-news/sandisk-slotmusic)
[35] "slotRadio" (http://www.sandisk.com/consumer-products/slotradio). SanDisk. . Retrieved 2011-11-27.
[36] SanDisk Ultra II SD Plus USB/SD card (http://www.theregister.co.uk/2005/07/25/review_sandisk_ultra_ii_sd_plus/) The Register, 2005-07-25
[37] "A-DATA Super Info SD Card 512MB" (http://www.techpowerup.com/reviews/AData/ADATASuperInfoSD/). techpowerup.com. 2007-02-20. . Retrieved 2011-12-30.
[38] "SD Part 1, Physical Layer Simplified Specification, Version 3.01" (http://www.sdcard.org/downloads/pls/simplified_specs/Part_1_Physical_Layer_Simplified_Specification_Ver_3.01_Final_100518.pdf). . Retrieved 2011-12-04.
[39] Fragmentation and Speed (https://www.sdcard.org/developers/overview/speed_class/) SDCard.org
[40] "SD Speed Class/UHS Speed Class" (https://www.sdcard.org/developers/overview/speed_class/). . Retrieved 21 Nov 2011.
[41] "Kingston Technology Company - Flash Memory Cards and X-Speed Ratings" (http://www.kingston.com/flash/x/default.asp). Kingston.com. . Retrieved 2010-08-22.
[42] "SD cards branded with an upper-case 'I' are faster, yo" (http://www.engadget.com/2010/06/24/sd-cards-branded-with-an-upper-case-i-are-faster-yo/). Engadget. . Retrieved 2010-08-22.
[43] "SD Card Association announces UHS-II, ultra high-speed SD card specification" (http://www.robgalbraith.com/bins/content_page.asp?cid=7-11133-11156). Rob Galbraith. 2011-01-05. . Retrieved 2011-01-05.
[44] MEAD-SD01 SDHC card adapter (Sony) (http://pro.sony.com/bbsc/ssr/micro-xdcamexsite/cat-accessories/product-MEADSD01/)
[45] "TS-7800 Embedded" (http://www.embeddedarm.com/products/board-detail.php?product=TS-7800). Embeddedarm.com. . Retrieved 2010-08-22.
[46] "Embedded SD" (http://www.sdcard.org/developers/overview/embedded_sd/). SD Card Association. . Retrieved 2011-11-30.
[47] "iNAND Embedded Flash Drives" (http://www.sandisk.com/business-solutions/inand-embedded-flash-drives). SanDisk. . Retrieved 2011-11-30.
[48] "Linksys WRT54G-TM SD/MMC mod - DD-WRT Wiki" (http://www.dd-wrt.com/wiki/index.php/Linksys_WRT54G-TM_SD/MMC_mod). Dd-wrt.com. 2010-02-22. . Retrieved 2010-08-22.
[49] "Press Releases 17 July 2003" (http://www.toshiba.co.jp/about/press/2003_07/pr1701.htm). Toshiba. 2003-07-17. . Retrieved 2010-08-22.
[50] SanDisk Introduces The World's Smallest Removable Flash Card For Mobile Phones-The miniSD Card (http://www.sandisk.com/Corporate/PressRoom/PressReleases/PressRelease.aspx?ID=1536) SanDisk.com
[51] PhoneScoop - Sandisk T-Flash announcement. (http://www.phonescoop.com/news/item.php?n=801)
[52] SanDisk Introduces 4GB miniSDHC Flash Card for Mobile Phones (http://www.sandisk.com/Corporate/PressRoom/PressReleases/PressRelease.aspx?ID=3530) SanDisk.com
[53] "Sharp Linux PDA promotes the use of proprietary SD card, but more open MMC works just fine" (http://www.linux.com/archive/feed/20060). Linux.com. . Retrieved 2010-08-22.
[54] Simplified Specification Agreement (http://www.sdcard.org/developers/tech/sdcard/pls/) from the SDA's website
[55] "OLPC mailing list archive" (http://mailman.laptop.org/pipermail/community-news/2006-September/000023.html). Mailman.laptop.org. . Retrieved 2010-08-22.
[56] "Simplified Version of SDIO CARD SPEC" (https://www.sdcard.org/developers/overview/sdio/sdio_spec/). SD Card Association. . Retrieved 2011-12-09.
[57] "SDXC memory cards promise 2TB of storage, 300MBps transfer" (http://www.engadget.com/2009/01/07/sdxc-memory-cards-promise-2tb-of-storage-300mbps-transfer/). Engadget.com. . Retrieved 2010-08-22.
[58] "Optimizing Linux with cheap flash drives" (http://lwn.net/Articles/428584/). Linux Weekly News. . Retrieved 2011-04-11.
[59] microSD & microSDHC Cards (http://www.toshiba-memory.com/en/micro_sd_cards.html) TOSHIBA Memory Solutions
[60] A look into how SDHC will affect the future Nand Flash market (http://www.dramexchange.com/WeeklyResearch/Post/1/492.aspx). DRAMeXchange, December 2006
[61] SD Compatibility (http://www.hjreggel.net/cardspeed/special-sd.html), CARDSPEED - Card Readers and Memory Cards, December 1, 2006
[62] "WinXP SP3 can't read 4GB SD card in multicard reader" (http://www.eggheadcafe.com/software/aspnet/33338344/winxp-sp3-cant-read-4gb.aspx). Eggheadcafe.com. . Retrieved 2010-08-22.
[63] JEDEC MMC 4.4 Standard Pg.7 (http://www.jedec.org/download/search/JESD84-A44.pdf), http://www.jedec.org 2008
[64] Transcend v4.0 card does not support 1.8V (http://www.transcendusa.com/support/dlcenter/datasheet/TSxxMMC4.pdf), http://www.transcendusa.com 2009

External links

Organizations

- Official website (http://www.sdcard.org), SD Card Association, sdcard.org
  - Membership: $2000/yr for General, $4500/yr for Executive.
  - Full Specification: Free for members, $1000/yr for R&D non-members.

Specifications

- SD Simplified Specifications (https://www.sdcard.org/downloads/pls/), sdcard.org, Free
  - Part 1, Physical Layer, v3.01
  - Part A1, ASSD Extension, v2.00
  - Part A2, SD Host Controller, v3.00
  - Part E1, SDIO, v3.00
  - Part E2, SDIO Bluetooth Type-A, v1.00
- Microsoft Extensible Firmware Initiative FAT32 File System Specification (http://download.microsoft.com/download/1/6/1/161ba512-40e2-4cc9-843a-923143f3456c/fatgen103.doc), Microsoft

Software

- SD Formatter for SD / SDHC / SDXC cards (Windows and Mac) (https://www.sdcard.org/downloads/formatter_3/), sdcard.org

Comparisons

- Comparison of numerous memory cards and readers, plus technical information (http://www.hjreggel.net/cardspeed/index.html#special-sd.html), hjreggel.net
- SDHC Card comparison (german) (http://www.hardware-infos.com/tests.php?test=81), hardware-infos.com
- Speed Testing of UHS-1 Cards in prosumer Nikon D7000 Camera (http://glamourphotography.co/gear/uhs-speed-class-1-sdhc-memory-cards-tested-delkin-8gb-elite-633x-secure-digital-uhs-i-95-mbsec-vs-sandisk-extreme-pr)
- Nikon D7000 Memory speed tests, SDHC Memory Cards: including Sandisk Extreme Pro UHS Speed Class 1 45 Mbyte/s and Several Popular Class 10 SDHC Cards (http://glamourphotography.co/gear/nikon-d7000-memory-speed-tests-sdhc-memory-cards-including-sandisk-extreme-pro-uhs-speed-class-1-45mbsec-and-se)

Interfacing

- Interfacing to SD cards, great technical details (http://elm-chan.org/docs/mmc/mmc_e.html)
- Interfacing AVR (Arduino) to SD cards, C source code (http://www.dharmanitech.com/2009/01/sd-card-interfacing-with-atmega8-fat32.html)
- Interfacing ARM to SD cards, C source code (http://gandalf.arubi.uni-kl.de/avr_projects/arm_projects/arm_memcards/index.html)
- Interfacing MSP430 to SD cards, C source code (http://www.cs.ucr.edu/~amitra/sdcard/Additional/sdcard_appnote_foust.pdf), Michigan State University
- Interfacing MAXQ2000 to SD cards, good technical descriptions, C source code (http://www.maxim-ic.com/appnotes.cfm/an_pk/3969), maxim-ic.com
- SD card controller, Verilog source code (http://www.opencores.org/project,sdcard_mass_storage_controller), opencores.org

ActiveSync

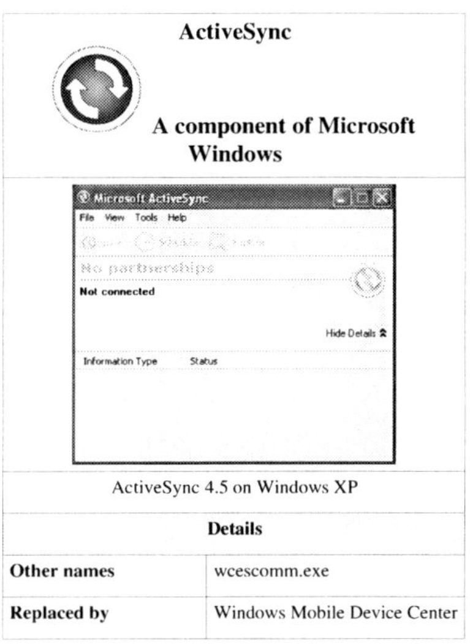

ActiveSync

A component of Microsoft Windows

ActiveSync 4.5 on Windows XP

Details	
Other names	wcescomm.exe
Replaced by	Windows Mobile Device Center

Initial release	1.0 / September 10, 1996
Stable release	4.5 / February 13, 2007
License	EULA
Website	microsoft.com [1]

ActiveSync is a mobile data synchronization technology and protocol developed by Microsoft, originally released in 1996. There are two implementations of the technology: one which synchronizes data and information with handheld devices with a specific desktop computer (originally known as *Handheld PC Explorer*), and another technology, commonly known as **Exchange ActiveSync** (or *EAS*), which provides push synchronization of contacts, calendars, tasks, and email between ActiveSync-enabled servers and devices.

Exchange ActiveSync is a proprietary protocol and is licensed to a number of mobile device companies, including Apple for iOS, Palm for its webOS devices, and Google for certain Android smartphones. Exchange ActiveSync technology is also used by other messaging and collaborative software servers, including Novell GroupWise and Lotus Domino and FuseMetrix. In the Windows Task Manager, the associated process is called **wcescomm.exe**.

Desktop ActiveSync

The desktop ActiveSync program allows a mobile device to be synchronized with either a desktop PC, or a server running Microsoft Exchange Server, Atmail, Axigen Mail Server, Horde, PostPath Email and Collaboration Server, Critical Path Messaging Server, Kerio Connect, Scalix, Zimbra or Z-push. Only Personal information manager (PIM) data (Email/Calendar/Contacts) may be synchronized with the Exchange Server. (Tasks may also be synchronized with Exchange Server on Windows Mobile 5.0 devices.) The PC synchronization option, however, allows PIM synchronization with Microsoft Outlook, along with Internet "favorites", files, and tasks, amongst other data types. ActiveSync doesn't support all features of Outlook. For instance, contacts grouped into subfolders aren't transferred. Only the contacts which are not in a subfolder are synchronized.

ActiveSync also provides for the manual transfer of files to a mobile device, along with limited backup/restore functionality, and the ability to install and uninstall mobile device applications.

Supported mobile devices include PDAs or Smartphones running Windows Mobile, or the Windows CE operating system, along with devices that don't use a Microsoft operating system,[2] such as the iPhone, Palm OS and Symbian platforms. Windows Phone 7 doesn't support desktop ActiveSync synchronization[3].

Starting with Windows Vista, ActiveSync has been replaced with the Windows Mobile Device Center, which is included as part of the operating system.[4]

Exchange ActiveSync

In addition to the ActiveSync desktop sync software bundled with Windows, Microsoft also uses the ActiveSync name to refer to the push messaging component of Exchange Server called Exchange ActiveSync, which relays messages to mobile devices.[5]

Release history

Version	Operating systems	Release date	Major changes
1.0	Windows 95	1996-09-10	• Initial version (under name H/PC Explorer)
1.1.7077	Windows 95, NT 4	1997-03-19	• NT4 support • Stability and compatibility fixes • Revised EULA
2.0		Fall 1997	• Renamed to Windows CE Services • Support for Windows CE 2 • Windows CE 1 support dropped
2.1		1998-02	• Support for Windows CE 2.1x
2.2	Windows 95, NT 4, 98	1998-09	• NT4 installation easier • 33% faster than 2.1
3.0.0.9204		1999-08-16	• Renamed to ActiveSync • Faster, simplified, and vastly improved. • Removed the association between RAS/DUN and the Windows CE connection stack
3.1.9386		1999-11-24	• USB synchronization and the inclusion of the AvantGo host client • Auto-adjusting baud rate
3.1.9439	Windows 95, NT 4, 98, 2000	?	• Sync fixes
3.1.9587		2001-07-31	• Added synchronisation support for Microsoft Exchange Server 2000 • Fixes for Outlook 98 / 2000 Security updates • Fixed a problem with the USB sync option
3.5.1176	Windows 95, NT 4, 98, 2000, XP	2001-08-06	• Integrated support for new Windows XP and Office XP releases and Pocket PC 2002 • Improved USB functionality, security and sync performance • New connection sounds
3.5.12007		2002-03-01	• Revised high color program icon

3.6.2148	Windows 95, NT 4, 98, 2000, XP, 2003, Home Server	2002-11	• Support for the new range of Smartphone devices • Customary security updates and synchronisation performance improvements • New Get Connected Wizard • Improves on remote synchronisation by preventing error messages and dialogues from halting sync process
3.7.3083		2003-05-06	• Minor updates to internal icon set • Corrects a discovered security flaw in ActiveSync • Improvements to the synchronisation wizard, other UI changes and general enhancements • Improvements to support forthcoming Microsoft Office 2003 release and Windows Mobile 2003
3.7.1.3244		2003-10-10	• Improvements to USB drivers and issues related to synchronisation • Get Connected Wizard's interface modified slightly.
3.7.1.4034		2004-03-26	• Fixed reported bugs with Windows Explorer and XP Firewall integration.
3.8		2005-01-06	• Secure functionality and provide updates for Windows XP SP2 systems • Performance improvements in synchronisation • Circumvents XP Firewall prompts that users experienced with other program versions upon first run. • Disables the Ethernet (LAN, Bluetooth) and RAS (Modem and WAN) connection method by default
4.0.4343	Windows 2000, XP, 2003, Home Server	2005	• Users able to specify installation directory • Removal of on-personal area connectivity options from the synchronisation mix. • Services for connections with Microsoft SQL server are included, along with a synchronisation update for Windows Media Player 10 • GUI refresh
4.0.4358		2005	• Retail version included only on Windows Mobile 5 device CDs
4.1.0.4841		2005-11-18	• Critical update
4.2.0.4876		2006-06-06	• Microsoft Outlook improvements: Resolves issues relating to error code 85010014 • Proxy/DTPT interaction improvements: Improved auto configuration of device Connection Manager settings when desktop has no proxy path to the Internet • Improved Desktop Pass Thru behavior with ISA proxy failures • Partnership improvements: Better resolution of multiple devices with the same name syncing with the same desktop • Connectivity improvements: Better handling of VPN clients (resolve unbinding of protocols from our RNDIS adapter). New auto detection of connectivity failure with user diagnostic alerts • New troubleshooting utility
4.5.5096		2007-02-13	• Faster file transfer speed and photo sync via Outlook are only available for Windows Mobile 5.0 powered devices. • Users of Microsoft Exchange 2003 Service Pack 2 with devices running the Messaging and Security Feature Pack for Windows Mobile 5.0 will benefit from the following feature enhancements included in ActiveSync 4.5: Direct Push Technology, local device wipe, and certificate powered authentication to Microsoft Exchange. • Microsoft Office Outlook 2000 not supported • Conversion of database files for use on a mobile device is not supported • Conversion of font files for use on a mobile device is not supported by ActiveSync 4.5

References

[1] http://www.microsoft.com/windowsmobile/activesync/default.mspx
[2] Microsoft Exchange Server: Exchange Server 2007 Support for Mobile Devices (http://www.microsoft.com/exchange/evaluation/features/owa_mobile.mspx)
[3] http://pocketnow.com/windows-phone/how-to-sync-windows-phone-7-with-outlook
[4] download.microsoft.com (http://web.archive.org/web/20070318042657/http://download.microsoft.com/download/7/0/9/70964f31-bac7-4379-b8bf-17ef35301ace/MEDC/Windows+Mobile+Enterprise+Features+-+Derek+Snyder.pdf)
[5] "Exchange ActiveSync: Frequently Asked Questions" (http://technet.microsoft.com/en-us/exchange/bb288524.aspx). Microsoft. .

See also

- Push email
- Handheld PC
- Handheld PC Explorer
- Palm-size PC
- Pocket PC
- Smartphone
- SyncToy
- Windows Mobile Device Center
- SynCE An Open Source Alternative

External links

- Microsoft ActiveSync (http://www.microsoft.com/windowsmobile/activesync/default.mspx)

Article Sources and Contributors

Nokia_E50 *Source*: http://en.wikipedia.org/w/index.php?title=Nokia_E50 *Contributors*: Acalamari, Apalsola, Apocalypso™, Armando, Asimzb, Bentogoa, Chowbok, Denisarona, Dgies, Evert, Fatla00, Fiftyquid, G patkar, Ghettoblaster, Godsofchaos, Gueneverey, JadeH, Kunz506, Ligulem, MMuzammils, Manish.parmar.b, Misi91, MrOllie, Muhandes, Nakon, Notmicro, Ohconfucius, Orangeaurochs, Pb30, Polluks, Raunzoss, Scientizzle, Searchmaven, Skype in Nokia, Smooth O, TastyPoutine, Tazfonebiz, TigerK 69, Typ932, Uditsood, Veikk0.ma, Wibbble, ZeroUm, 35 anonymous edits

Nokia *Source*: http://en.wikipedia.org/w/index.php?title=Nokia *Contributors*: -Majestic-, 007zoo, 130.236.221.xxx, 159753, 16@r, 1exec1, 205ywmpq, 28421u2232nfenfcenc, A bit iffy, A box of sticks, A. B., A333, AB, ABF, AEMoreira042281, Abdul raja, Abnyc81, Academic Challenger, Accurate Nuanced Clear, Acdx, Acela Express, Adamrush, Adaobi, Adashiel, Adraeus, Aesopos, Aeusoes1, Agusm266, Ahazred8, Ahmedcena, Ahoerstemeier, Aijoovai, Aintneo, Alansohn, Aldis90, Alepik, Alex43223, Alexius08, AlfredWalsh, AliShaikh85, Alirezazzz, Alisha.4m, Allanvs, Allpower, Amitn, Andres, AndrewHowse, Andros 1337, Andyabides, Anole 418, AnonMoos, Antandrus, Antoncampos, Anubhavsharmaa, Apalsola, ApnAEA, Apparition11, Arch dude, Ariele, Armando, Arthena, Arungm29, Arunsingh16, Asanka000, Ashmoo, Ashutosh.manager007, Atanasov, Avoided, Axeman89, BLAZEXBOY, Backwalker, Badr55, Barek, Batbayarl, Beaker1306, Beao, Bearcat, Beetstra, Benhocking, Bfaabaa, Big Brother 1984, BigT333, Billyboy1980, Blake-, Blakegripling ph, Blobglob, Bloggert, Bluezy, Bmannaa, Bobblewik, Bobo192, Bogdangiusca, Bongomatic, Bongwarrior, Boomshadow, BorisxXx, BoyanSyarov, Brennish, Brockert, BrokenSegue, Brokestudent007, Bruce1ee, BunnyT, C. A. Russell, C.Fred, C311u1ar, C628, CONFIQ, Cablecord, Cameron Scott, Can't sleep, clown will eat me, CanadianLinuxUser, Canterbury Tail, Capricorn42, Catiana 465, CecilWard, Cerebrith, Chamal N, Cheezy man, Cherkash, Cheung1303, Choptube, Chris the speller, ChrisHodgesUK, Chrisissocool, Clarkedexter, ClementSeveillac, CliffC, Clipmode, Closedmouth, Cmreditor, Cntras, Codey123, Coguar, Colibri37, Conversion script, Coolbull20, Coolslko, Cpl Syx, Crevox, Cryonic07, Cybercobra, Dale Arnett, Damiens.rf, Damirgraffiti, Dancter, Daniel*D, Daniel575, DanielCD, Danim, Danio, DarkSaber2k, Darth Panda, Davidbspalding, Dck7777, Dearsina, Debresser, Delta avi delta, Den fjättrade ankan, Desaivishal14, Desbiadi, Diasimon2003, Dicklyon, Dicostathomas, Dima1, Discospinster, Dissolve, Diyar se, Djr xi, DmitTrix, DocendoDiscimus, Doctorcasey, DomQ, Dostal, Dostick, Download, Drewt, Drpickem, Dsavi.x4, Dyl, E Pluribus Anthony, E000xm, ES Vic, EWikist, Ed g2s, Editor182, Edsuom, Educatednawab, Edward, Eekerz, Eenu, Elfguy, Eliz81, Emmalewis1, Enemenemu, Epoxed, Eraserhead1, Esebi95, Estoy Aquí, Etincelles, Eurocanna, Everyking, FR Soliloquy, Falcon8765, Falconoffrance63, Feezo, Felix Dance, Fffaaattt, Fieldday-sunday, Fixer88, FleetCommand, Flewis, Flora, Florentino floro, Florin92, Flrn, Flyguy649, Folksong, Fooishbar, Force39, Fram, France64160, Franz-kafka, Fredrik, FreplySpang, Frymaster, FunkyDuffy, GMRE, Gachen, Galeon54, Galoubet, Gambit 28, Gandhietami, Gaurav13dubey, Gavinio, Gdo01, Georgy90, Gethresh, Ghodannywahyudi99, Gimboid13, Gnangarra, Gnuton, Gobonobo, Godospoons, Gogo Dodo, Golemeye, Gr1st, Grafen, Graham87, Gratom, GreenJellybean, GreenJoe, Greenshed, Gregorydavid, Gringer, Groshna, Gsarwa, Guaka, Gump Stump, Gunglewack, Gunmetal Angel, Guy M, H.lloyd, H0dd0ck, Haakon, Hadal, Hankwang, Harishua, Harmi banik111, Harriv, Hatesonyericsson, Hauskalainen, Havarhen, Hdt83, Head, Hell9, HelloAnnyong, Hemanshu, Heron, Herr Beethoven, HkCaGu, Hmains, Hooperbloob, Hotwiki, Howardchu, Htanna, Hu12, Hustedcarl, Hydrargyrum, Hylene, IJK Principle, Icewindfiresnow, Icseaturtles, Ilovemymac, Imgaril, Immunmotbluescreen, Imperi, Inspirito, Intelligentsium, Interframe, Inzy, Iohannes Animosus, Iphon, Irishnbears, Irstu, Isfisk, Ithinkicrappedmyself, Ixfd64, J.delanoy, J36miles, JCDenton2052, JForget, JGRIFF47, JIP, JLaTondre, JNW, Ja 62, Jackollie, Jak123, Jamcib, JayceAndTheNews, Jbsegal, Jclemens, Jdl18, Jeffrey Mall, Jeltz, Jensbn, Jeronimo, Jerry, Jerryseinfeld, Jim, Jimp, JmeSaunders, Jmh, Jni, JoeHinks, Joel7687, John, John Appleseed, JonForst6, Jonathan Hall, Jonik, Jopo, JorgeGG, Joseph Solis in Australia, Jovianeye, Jpk, Jrdioko, Jsysinc, JudasJesus, Justin Steele, KC., KFP, KUsam, Kabsingh, Kai Ojima, Kalivd, Kallemax, Kalpesh.v.mistry, Kangaroopower, Katous1978, Kcandrsn, Keegan, Keonne, Ketchup, Khanri01, Kickus, Kirosana, Kjetilho, Kjramesh, Kkm010, Klilidiplomus, KoastalBenefitPromo, Konakalla anurag, Konrad Foerstner, Krellis, Kukulcan 7560, Kuponadam, Kwamikagami, LG4761, LLentil, Lafraia, Lamat, Larry laptop, LarryGilbert, Ld100, Leandrod, LeaveSleaves, Lectonar, Leo-loErlahell, Lepensky, Levineps, Lewys, Lfh, Lhotkami, Liamgilmartin, Lightmouse, LilHelpa, LittleDan, Loigenth, Lollomama, Look2See1, Loren.wilton, Lovesameer9812, Lowflyingowl, Luigiacruz, Luisvmejia, Lukobe, M3lm4tt, MER-C, MMuzammils, Ma.abilash, Mabdul, Magioladitis, Makedonec28, Makeemlighter, Malhonen, Mani1, Manop, Marc Lacoste, Mardus, Marek69, Mark Ryan, Marmzok, Martarius, Mass09, Mastermind 147, Matrobriva, Matt povey, Mattbr, Mav, Maxim4o, Mayurg, McSly, McWika, Megahmad, Menphrad, Mert2000, MetroStar, Mhkay, Mic, Michaelbarr123, Microtony, Mike Rosoft, Mimsie, Minimac, Miraceti, Mirmo!, Misspsyb2, MithrandirAgain, Mkidson, Mmsteelers, Mojei, MominS, Momirt, MonaNL, Monkeynoze, Mowsbury, Mr Stephen, Mr. Met 13, MrFawwaz, MrOllie, MrZoolook, MrsAmethystWay, Mumble45, Murphy418, Mysdaao, NKDurrani, Nagytibi, Nakon, Nantasatria, Narcisso, NawlinWiki, NeilN, Neilgravir, Neon white, Newone, Newtechpulse, Nielswik, Ninja5624, Ninjustic, NokiaF, Nokiatrader, Noraft, Norden83, Nscheffey, Nubbly, Nubiatech, Number29, Nuno Tavares, Nuttycoconut, Nwpl, Obli, Ohnoitsjamie, OkakiMCMLXX, Oleksandr Kononenko, Oli Filth, Olivier, Omernos, One, Onlynokia, Onorem, Oo64eva, Ooza, Opraco, Ostralek, Otto2011, Oxymoron83, P0ppe, PTSE, Pablo-flores, Palefire, Parthrana, Pascal.Tesson, Pavel Vozenilek, PeeJay2K3, Pengo, Perohanych, Persia2, Pessi, PeteS, Peter Chastain, Petri Krohn, Pgan002, Pgk, Philip Trueman, Phonefinder2007, Pink Bull, Pista235, Pkadam, PkerUNO, Ploca12, Pmggp, Pne, Pol098, Pony1142, Ppntori, Prakash.Akshat87, Prari, ProhibitOnions, Prokopov, Prolog, Proudfoot 001, Pseudomonas, Pterre, Puckly, Pudeo, Puneeeetjain, Pupster21, Quackdave, Quattrope, QuiteUnusual, Qwyrxian, Qxz, R'n'B, RTG, RVJ, Radagast83, RadicalBender, Ragityman, Rangoon11, Ratinator, Ratul655, Ravi vadgama, Ravo492, Rcawsey, RedWolf, Reliancepowercoin, Rent A Troop, Retired user 0001, Rettetast, RexNL, Rhobite, Richard Harvey, Ricktherazor, Riklear, Rjwilmsi, Rklawton, Roamataa, Rob Lindsey, Rob1974, Ron Ritzman, Ronnotel, Rrjanbiah, Rune X2, Ruronimomo, Ruslan0202, S h i v a (Visnu), S3000, SMC, SMP, ST47, Sadunlove, Sai2020, Saimhe, Salilm, Salome5764, Sam Hocevar, Samjuise, Samsara, Sander Säde, Sanixia, Sanjiv swarup, Satchmo2010, Savh, Savvo, Scootey, ScottSteiner, Scupplefish, Seanmilloy, Sebastian Shaw 449, Secaundis, Sedathut, Selimbey, Seqsea, Sfan00 IMG, Sfmammamia, Shadowjams, Sharcho, Shashankbhat, Shaun680, Shawnnicholsonca, Shd, Sherool, Shikker, Shizane, Siafu, SidP, Silpol, Silvah 87, SimonCrowley, SimonThird, Simone, Sinigagl, Skittle, Skyezx, Slavon37, Slo-mo, SmartFace44, SnappingTurtle, Snowolf, Sokratees9, Someguy1221, Soupyjnr, Spangineer, SpigotMap, Squash Racket, Squirtypants, Starblind, Steel, Stephan Leeds, Stephenb, SteveDay, SteveSims, Studioghiblitotoro, Suhailpeerbhai, Suomi Finland 2009, SuperHamster, Sven Manguard, T.O. Rainy Day, THEunique, TYelliot, Tagishsimon, TarzanASG, Tascha96, Tbhotch, Tda, Tedats, Teksosyete, Tellarin, Templetongore, Teqhed, Term061, Test2010, The Man in Question, The Person Who Is Strange, The Random Editor, The Rogue Penguin, The Thing That Should Not Be, TheBlueKnight, TheGreenFaerae, TheYmode, Thebluebeast, Themfromspace, Thingg, Thorvy, Thule, Thunderbrand, TicketMan, Tide rolls, TigerK 69, Timo Honkasalo, Timonoko, Timster69, Tkynerd, Toehead2001, TomB123, Tomisti, Tompagenet, Tomwalden, Tonius, Tony1, Topperfalkon, Tpbradbury, Trakesht, Treschupetes, Tri400, Triage, Trisreed, Tsungik, Tuliopa, Turnstep, Twid, Ufinne, Ulric1313, Ultraviolet scissor flame, Ulysses, Uncle Dick, Unclealex, UpBeat, UpstateNYer, Valentine McKee, Varundbest10, Velella, Venus 9274, Versageek, Vespristiano, Vianello, Vina, Vinayakgole, Vinaywin7, Vininche, Vitund, Vkem, Volgar, Vrenator, Vuo, WIMYV, Wahgujarat, Waycool27, Weatherwax, Welovedoves, Weyes, Wfaulk, Wibbble, Wiki Wonda, Wikipelli, Wikizard2010, Wikizeta, William M. Connolley, WilliamKF, Wimt, WiseOne76, XLerate, Xhienne, Xyzahirm, Yakudza, Yamaaan, Yamamoto Ichiro, Yandman, Yanksox, YellowMonkey, Yousaf465, Yuhani, Yvresgyros, Z10x, ZZninepluralZalpha, Zackwee, Zeno Gantner, Zidonuke, ZirconiumTwice, Zodiak 887, Zpetro, Zsinj, Zunils, דוד55, عبدالرحمل ادمحم ارمان محد, ஹெலெசின்கி பேனாந்த, , 1308 anonymous edits

Nokia_Eseries *Source*: http://en.wikipedia.org/w/index.php?title=Nokia_Eseries *Contributors*: -Majestic-, 8th Gamer, Adeeb01, Andries, Armbrust, Asimzb, Beyoncé Superfanatic, Bhimaji, Boban23, Borgx, Bv abhi, CommonsDelinker, Darkfire07, Dermotmallon, Dicklyon, Djr xi, Ed g2s, Einsidler, Eleman, Fazlani, Feathered serpent, Fiftyquid, Georgy90, Hu12, Irfan2009, JHeinonen, Jack007, K7aay, Kanonkas, Kyrios320, LaMenta3, Ligulem, MMuzammils, Mackrauss, Mafweb, Marc Lacoste, Maxim Masiutin, MrJacky, Muro de Aguas, Nixeagle, Notmicro, OwenBlacker, Pbxnsip, Perohanych, Prolog, R'n'B, Rich Farmbrough, Rockysmile11, Rror, Sapibobo, Saxphile, Shii, Stepa, Subsume, Taras, Tarc, TigerK 69, TinyFirstman, Towel401, Varun.c.jain, Vaskopopov, Virgi44, Wibbble, Xaje, 87 anonymous edits

Multi-band *Source*: http://en.wikipedia.org/w/index.php?title=Multi-band *Contributors*: Bearcat, Flowzn, GJDR, Hankwang, Lkt1126, Llorenzi, Pinethicket, R'n'B, Rchandra, Sdrazfar, TerriersFan, Vitorvicentevalente, Wtshymanski, Wusel007, 8 anonymous edits

Intellisync *Source*: http://en.wikipedia.org/w/index.php?title=Intellisync *Contributors*: ArrowmanCoder, Belino, Can't sleep, clown will eat me, Cander0000, Chirags, Cmreigrut, Dsouza, Edward, FleetCommand, Fritz Saalfeld, Gentgeen, Gillyweed, Grafen, Hmbr, Illyria05, JLaTondre, Josh Parris, NotWith, Ottawahitech, Scruzin, Stepheng3, The wub, Tophers, Tsange, Wamra, Wellspring, 13 anonymous edits

Smartphone *Source*: http://en.wikipedia.org/w/index.php?title=Smartphone *Contributors*: 16@r, 3Coins, A Man In Black, A123a10, A8UDI, AKA MBG, ALSLUG, AVM, Aaditya 7, Acalamari, Acather96, Aclassifier, Acolin f, Adriatikus, Aeons, Ahoerstemeier, Aillema, Akuyume, Alansohn, Alex890, Alf Boggis, Alf.laylah.wa.laylah, AlistairMcMillan, Allen3, Amarendra.Avinash, AndrewSpec, Andries, Andros 1337, Angela, AniRaptor2001, Ansible, Anupam, Arab Dynasty, Arafael, Arny, Arp120, Arunsingh16, Atama, Atenyi, Atul.ecn, Avoided, BD2412, Bahaltener, Bakshi41c, Balazer, Baronnet, Beland, Bendes, Bhny, Bibijee, Biker Biker, BillHaywood, BlkStarr, Bomazi, Bovineone, Branddobbe, Brianreading, Brittanybee1121, Broccoli, CLW, CRJO-CRJO, Cactusframe, Cadiomals, Caj27, Calabe1992, Calcwiki, CalumCook234, CanadianPenguin, Canalteen, CatherineMunro, Cellcom, ChaChaFut, Changshui88, Charivari, Chris the speller, Ciphers, Citizensmith, Cleared as filed, Closedmouth, Clovis Sangrail, Cmf, Cmlewan, Cmp101, Codetiger, Comaleaf, CommonsDelinker, Commontimect, Cool Hand Luke, Coolaaron88, CoolingGibbon, Cornea503, Cst17, Currab, Cxtom, Cyruslei, DaisyField, Dale Arnett, Dan6hell66, Daniel.finnan, DaveChild, Deljr, Dcxf, Ddavid2005, Deane@gooroos.com, Delphii, DerBorg, Deviceapps, Diannaa, Diego Moya, DivineAlpha, Dmarquard, Dreambringer, Drobatch, E.au, EVula, EWikist, Editor182, Editrrr, Elassint, Electronicbazaarnz, Electronicguru1, Emma23 K, Epolk, Eraserhead1, Erc, Erianna, Eshade, EugeneZelenko, Eugenia loli, Evice, Evilruletheworld96, Evolutiondb, F1MotoGPWRC, FCYTravis, Farolif, Father Goose, Fcassia, Feinoha, Feydey, Fhontoy, Flowanda, Fourdee, Furrybeagle, Fuzheado, GTBacchus, Galadh, Garant^ ^, GateKeeper, Geologyguy, Georgy90, Gfannick, Giftlite, Giggett, Gilliam, Ginsengbomb, Giraffedata, Glacialfox, Glennwells, Gmumru, Gogo Dodo, GoingBatty, Gold Hat, Gomm, Gookey, Gordon Ecker, GraemeL, Grahamperrin, Grajales, Graywolf, Graywolfmoon1, Gregory Heffley, Gregorysalt, Greswik, Groink, Gsarwa, Guy Harris, Gwernol, HDCase, Ha us 70, Haakon, Hadiceberg, Halloween.mac, Hamiltha, Hammersoft, Hardylane, Harmil, Harryzilber, Haseo9999, Hede2000, Henrik, HereToHelp, Hervegirod, Hobartimus, Hu12, Human.v2.0, Huntersquid, HuskyMoon, HybridBoy, Hydrogen Iodide, Hydrox, Hypnotist uk, I already forgot, I, Podius, Ianb, Ianboudreault, Ianereed, Icedwater, Ignis Fatuus, Illegal Operation, Illyria05, Im Buff, Immunmotbluescreen, Imroy, Imsilly123, In2thats12, Indefatigable, Interframe, InternetMeme, Intershark, Ipsign, Irky, Ishdarian, ItsZippy, J, J.delanoy, JCDenton2052, JCrue, JNW, Jagdterrier, JamesWeb, JamieS93, Jantangring, Jeffq, Jekyllhide, Jerome Charles Potts, Jerryobject, Jfdwolff, Jim.henderson, Jimmy Bergmark, Jobin RV, JoeSmack, John of Reading, JonHarder, Jonabbey, Jondel, JonoP, Joseph Solis in Australia, Josh Parris, Jostake, Jsherwood0, Jsribeiro, Juan M. Gonzalez, Jusdafax, Just another editor, Justin Mauger, Jwojdylo, Jæs, Kajowi, Kariteh, KarmaGeddon, Katieh5584, Kbdank71, Kcomstock, Kellen`, KelleyCook, Kentynet, Kevin Dorner, Kingpin13, Kinst, Klokbaske, KnowBuddy, Koavf, Kozuch, Kraftlos, Kresp0, Kudpung, KungFuMonkey, Kuru, Kwaichi, LaVieEntiere, Lai888, Lambtron, Larkhill97,

Laurasmith70308, Lazulilasher, LeilaniLad, Lenin1991, Lester, Leszek Jańczuk, Leujohn, Level plus, Lewispb, Lifes g00d 561, Lilac Soul, Limequat, Little Mountain 5, Little Professor, Logan, Loser997, Lotje, LrdChaos, Lukan4ica, Lun Esex, Lun4tic, MER-C, MMuzammils, Mac, Mac John Concord, Magnus.de, Malik Shabazz, Mange01, Maniacgeorge, Manop, Manway, Marcisjustice, Marco.difresco, Marcus Qwertyus, Mark Kim, Mark91, Marksza, Marrowmonkey, Martarius, Masgatotkaca, Mathiastck, Maury Markowitz, Maxí, Mcduck, Mckote, Mdrejhon, Mdwh, Meehawl, Meelar, Memaster3, Mephistophelian, MetaManFromTomorrow, Miaow Miaow, Mikael Häggström, Mike Dillon, Mike Linksvayer, Mikehelms, Mild Bill Hiccup, MillenniumB, Minghong, Miquonranger03, Mix Bouda-Lycaon, Mlg07e, Moberg, ModWilliam, Monaarora84, Mononomic, Morphh, Mortense, Mosmof, Mr. Strong Bad, MrOllie, Mvjs, My76Strat, MySchizoBuddy, MyronAub, Myscrnnm, Nakakapagpabagabag, Nathan94124, Navacell, Nazrich, Ne0Freedom, Nealmcb, Neo Ogilvy, Neoarchon, Neoguy999115, Nerdeff, Netrapt, Nick Number, Nickolai 420, Night Gyr, Nightscream, Nitro.ajb, Nixdorf, Nneonneo, Nopetro, Nurg, Obey, Ohnoitsjamie, Oknazevad, OlavN, Old port, Oli Filth, Omegatron, Oontay, OpenToppedBus, P.Marlow, PS., PTSE, Parintachin, Patrick, Pats1, Patvac-chs, Paul 1953, Pbrown111, Pdahomepage, Petershank, Peyre, Phatalflaw, Phatom87, Phearson, Philip Trueman, Piano non troppo, Pingveno, Piotrek54321, Pluma, Pmarshal, Pmlineditor, Pol098, Pol430, Poor Yorick, Posix memalign, Prathameshsasane, Prillen, Pvanheus, Pyfan, Quebec99, Querencia, R'n'B, RA0808, Radiier, Radon210, Rafael.sp, Ramu50, RedWolf, Reliablesoft, Repetition, Res2216firestar, ReverseEngineered, Reyk, Rgreed, Rhobite, Riadlem, Rich Farmbrough, Richardguk, Richiekim, Riki, Rollins83, Ron2, Ronz, Ruchir257, Rursus, SF007, SPQRobin, Sadegh87, Sainath468, Samad120, Samwb123, Sandstein, Sapibobo, Sayid98, Schrödinger's Cake, ScottyBerg, Scrtcwlvl, Sdfisher, Sdrazfar, Sean13zz, Sebastian Mandrean, Sebindcruz, Secretsorry, Serg3d2, Sfacets, Sfan00 IMG, Sfsmartp, Shalroth, Shawnc, Sheehan, Signalhead, Siliconov, Singerdg1, SirJibby, Skatebiker, Skintigh, Skittle, Slitchfield, Smart1954, Smartphonespakistan, Smmgeek, Snigbrook, Soliloquial, SpaniardGR, Spellmaster, Starofwonder, Stephen B Streater, Stephen Turner, Stjson, SuperHamster, Surenkarapetyan, Suwatest, Swoof, TVS99, Taka76, Takamaxa, Taketa, Tangent1000, Taskinen, TastyPoutine, Tatterfly, Technopat, Tesi1700, Tfgbd, ThaWhistle, ThaddeusB, The Pink Oboe, The Thing That Should Not Be, The Wild Falcon, The-apathy, Theaveng, Thestick, Thiled, Thisma, Thumperward, Ticell, Tide rolls, Tikru8, TimSE, Tklaer, Tmuller2, Tobias Bergemann, Togopogo, Tokyogamer, Tomas.turek18, Tombomp, Tommy2010, Tommytentimes, Tony1, Toussaint, Tpbradbury, Trasz, Treekids, Tuju, UCLATre, UnrealG, Urod, Uzume, Vegaswikian, Verkinto, Verne Equinox, Versus22, Vkem, Vrenator, Waqas Hussain, Westie4321, WhisperToMe, Whkoh, WiZZLa, Wideangle, Wiki admi, Wikimhb, WikipedianYknOK, Wikipelli, Winged-stone, Wireless Buddy, Wknight94, Woohookitty, XGrape, Xajel, Xrobertcmx, Yean3d, Youxiarock, Yug, Yunshui, Zach Vega, Zanter, ZimZalaBim, ZipoB, Zpetro, 1112 anonymous edits

Visto *Source*: http://en.wikipedia.org/w/index.php?title=Visto *Contributors*: 45Factoid44, CST, Caront, Chairman S., Colonies Chris, Crystallina, DaKenSu, DukeST4, Edward, Evrik, Garyp01, GoingBatty, Greenshed, Ian Pitchford, Idumont7, JLaTondre, Mixte, MrOllie, Nowa, Oli Filth, One, Palaeologus, Rigel1, Shattered, Smart Fox, Spoon!, Stepheng3, Vlad, Wellspring, Zotel, 51 anonymous edits

Secure_Digital *Source*: http://en.wikipedia.org/w/index.php?title=Secure_Digital *Contributors*: 24frames, 81120906713, 99bluefoxx, 9allenride9, A Man In Black, AVRS, Aab298347927384, Adam Schloss, Adam2288, Adsp, Aeons, Agateller, Aikenware, Airplaneman, Akaustav, Alansohn, Aldie, AlexJ, Alexbuster500, Alexthekiwi, Allaun, Alphaman, Alvis, Amoiwangjy, Amram99, Andrevan, Andrew sh, AndrewKepert, Andrewpmk, Andru nl, Angela, Angerdan, Animehawaii, Appraiser, Aqn, Aribronstein, Armando, Arrowcatcher, Asim18, AssetBurned, Atanasov, Atompowered, Austinmurphy, Autoandragogist, Axcelis555, AzaToth, BENNYSOFT, BPM, Bastique, Bdelisle, Beao, Bearcat, Beeline23, Beland, Bergsten, Bholstege, Billgordon1099, Binba, Bjdehut, Blanchardb, BlindWanderer, Blorg, Bobblewik, Bobianite, Boboangel, Bomazi, Bookandcoffee, Bostwickenator, Brer vole, Brian Patrie, Brian-L, Briandgregory, Brianski, Brouhaha, Buster2058, Buuneko, C 1, Calcprogrammer1, Caleson, Calor, Can't sleep, clown will eat me, Canrocks, Carreon77, CaseyPenk, Caspertheghost, Catskul, Cdrum, CesarB, Chandu15, CharlotteWebb, Chealer, Chris the speller, ChrisCork, Chrisjj3, Christian75, Ckmac97, Climber22, Clorox, Cmdrjameson, Compellingelegance, Compuguy1088, Coneslayer, Consumed Crustacean, Cootiequits, Copysan, Corge, Coughinink, CrookedAsterisk, Csendesmark, Cyawman, Cybercobra, Cybertai, CyclePat, D0nj03, DMahalko, Dabomb87, Dah31, Dale Arnett, Damian Yerrick, Dan100, DanMS, Dancingonmice, Dancter, Dandv, Daniel Newby, Danielt998, Dankru, Darin-0, Darkxsun, Darrien, Daveswagon, David from Downunder, David spector, David.Monniaux, Davidswelt, Dawd, Deineka, Dennbruce, DervishD, Dhanashekar, Djm1279, Dlrohrer2003, Doc glasgow, DocWatson42, Doctorno, Dogcow, Drabant, Draconiator, Dreadstar, Drmies, Dustin Howett, Długosz, EJSawyer, Echtner, Ed Poor, Eftpotrm, Electricnet, Electron9, ElementFire, Elemesh, Elvey, Elwilke, EncMstr, Ente75, Erencexor, Ericzhang789, Esowteric, Eurleif, Evice, Excirial, Exearly, ExplicitImplicity, ExportRadical, FCartegnie, Fast healthy fish, Feedmecereal, Feydey, Flarn2006, FlashSheridan, Flasher, Fleminra, Flowerflower, Fmillour, Freddyzdead, Fresheneesz, Fudoreaper, Fullerene, Funandtrvl, Furrykef, Fvw, Gadol87, Gaselvin, Ge0rge, Geekosaurus, Gglockner, Ghaib, Glaw10, GodGell, Goosnarrggh, GraemeLeggett, Grafen, Graham87, GregorB, Grunt, HDCase, Haipa Doragon, Hankwang, Hansa11, Harlekeyn, Harvester, Hebrides, Henry W. Schmitt, Heron, Heybales, Hgrosser, Hoagg, Hooperbloob, HoserHead, Hurk87, IAmAI, IW.HG, IceHunter, Ichimonji10, IkonicDeath, Iliev, Impi, Intelliot, InternetMeme, Irayo, Ismouton, J.delanoy, JJLatWiki, Jachim, JackTinWNY, Jafeluv, Jaho, James JK, JamesAM, JamesTeterenko, Jdowland, Jeffq, Jerryobject, Jhsounds, Jidanni, Jim McKeeth, Jim.henderson, Jj2235, Jlslspam, Jnavas, Joaoplim, Jocunddus, Joe Sewell, John of Reading, JohnAHind, JohnAlexan, Johnteslade, Jordan Brown, JordoCo, Jory, Josi.ow, JudyJohn, Justin Ormont, Jvr725, Jw21, Kartano, Kaszeta, Kazrak, Kbdank71, Kiimmyluff, Kinema, Kingpin13, Kjohna, Klkmiooo, KnowBuddy, Koolman2, Korj.by, Kozuch, KryptoCleric, KsprayDad, Kungfujoe, Kungming2, Kusunose, Kvng, LM1987, Lamrock, Lavenderbunny, Lbecque, Le Déchaîné, LeoO3, Leszek Jańczuk, Liaocyed, Lightmouse, LilHelpa, Lindosland, Little Professor, Littleendian, Lord Nightmare, Lordsatri, LostLeviathan, LovesMacs, Lucasreddinger, MER-C, MMuzammils, Macrakis, MadisSepp, Manuishino, MarXidad, Marianocecowski, Mark Renier, Martarius, Matthiaspaul, Maury Markowitz, Maxim Leyenson, Maxí, Mboverload, Megazero, Memorysuppliers, Metallica10, Michael L. Kaufman, MichaelJanich, Microsofkid, Mihai Capotă, Mikus, Misterkillboy, Mju7nhy6, Mmj, Mnts, Modster, Moreati, Mortense, Moxfyre, MrBurns, Mram80, Mrdungx, Mrzaius, Msaunier, Mwarren us, Myke2020, N1truX, N5iln, NEMT, Nagualdesign, Nahuel.carvajal, Nasukaren, Neelix, Neilka, NerdyNSK, Neurolysis, Newzack, Nil Einne, Nintendude, Nixdorf, Nnan, Noir, Noq, Notmicro, Now3d, Ntsimp, Nulltransfer, Nurg, Nv8200p, OMA2k, Oahiyeel, Oe1kenobi, Offbeatcinema, Ojigiri, Oliverdl, Omeganian, Omegatron, Oxymoron83, Pabouk, Paranoid, PatrikR, Peace81, Peaceninja28, Peter Campbell, Peter S., Peterl, Peterrobbemond, Peyre, Photographerguy, Photoguy439, Phr, Pickmynose1999, PierreOssman, Pinkevin, Piper8, Placi1982, Plasmaroo, Pranjal87, ProhibitOnions, Psharrock, Psiddall, Quaeler, Queefycreatures, Ravedave, Ravenperch, Razor2988, Rcawsey, RealGrouchy, Reaperducer, Rebroad, RichardTector, Riddley, RingtailedFox, Rjc34, Rjquillin, Rjwilmsi, Rlcantwell, Rmsuperstar99, Rmunn, Rofl cawpters, Rohan Jayasekera, RoseTech, Rosenbluh, Rstoplabe14, Ruud Koot, SD Card Version 2.Wiki.0, SF88, SGBailey, SMD915, SPKirsch, SRG275, Saimhe, Saipraneethn, Saltmiser, Sam8, SanGatiche, Sandymac, Saravanants, Sasan.j, Saverworld2, Sbmeirow, SchnitzelMannGreek, Scientus, Scotty jasper2000, SeanAhern, Seaphoto, Sebadee, Sedimin, Sergei, Shadowjams, Shanes, SidP, Sin-man, Skathol, Skierpage, Skyykj, Slakr, Slashme, Slicing, Smack, Smileyborg, Smyth, Snolygoster, Speedevil, Spike-from-NH, Spikesagal, Stanleyivan, Starnestommy, Steinsomers, Stephan Leeds, StephenFerg, Stevage, Stuston, Sunny house, Surf243, Suriel1981, Suruena, Szzuk, TAC-3, TCav, Tacvek, Tamiledu, Tarquin, TedPavlic, The Giant Puffin, The Thing That Should Not Be, TheDoober, Themoment, ThevillagesmithE, Thewikipedian, Thunderbird2, Thunderpenguin, TimSE, Timharwoodx, Tintenfischlein, Tiredofscams, Tkgd2007, Tmansour, TobleRone, Todd Vierling, Toehead2001, Tomgibbons, Tone, Tony1, Tonyhawz, Totakeke423, Towel401, Tpbradbury, Triscal1990, Tristan Schmelcher, Twinmostech, U4ia74, UU, UberMan5000, UltraMagnus, Vdcappel, Versus22, Vexorg, Viggio, Viol8or, Vladsinger, VoxLuna, Waffle, Wammes Waggel, Wanderer099, Wbt92, Wbuch, Weezee, Wernher, WhiteDragon, Whitis, Wibbble, WikHead, WinTakeAll, Wine Guy, Wirbelwind, Wjejskenewr, Wjw0111, Wolfling, WriterHound, Wrlee, Wtshymanski, Wuhwuzdat, XP1, Xapplimatic, Xdddex, Xenon54, Xmhd, Xonicx, Xpclient, Yaniv Kunda, Yealout, Z hosen, Zarcillo, Zarenor, Zbrahead91, Zerpent, Ziga, Zodon, Zoicon5, Zsexdrcft, Zurotai, Zviangi, 1151 anonymous edits

ActiveSync *Source*: http://en.wikipedia.org/w/index.php?title=ActiveSync *Contributors*: 16@r, AKA MBG, Aadh, Abelson, Akhristov, Althepal, Amplitude101, BobTheMad, Brianreading, C J B Scholten, C:Amie, Cabernet, Cmdrjameson, Cwolfsheep, Cynical, Dale Lane, Davewho2, Dnjuls, Doniago, Dr. F.C. Turner, E71, Eivind, Evice, Exe, Fernando S. Aldado, Fvw, Ghettoblaster, Gmdoan, Golem, Gtvcyhc, Gunnar Wrobel, IVO, Iandotcom, Idumont7, Illyria05, Iuhkjhk87y678, JLaTondre, Jeepday, Jæs, Kjwu, Koweja, Kslotte, Larrymcp, Lightmouse, M4gnum0n, MobileGlick, Mocheeze, NOSaturn, Nudecline, OsamaK, Pgr94, Philofred, Pigsonthewing, Pol098, Rebroad, Retroneo, Rob010, Rogerd, RoyBoy, Shshme, SimonP, Smearp, Soumyasch, Stephenchou0722, Ta bu shi da yu, The Anome, Tho2468, TobyDZ, Tomeasy, Tuju, Tyhopho, Wagner51, Warren, WillT.Net, Xpclient, 131 anonymous edits

Image Sources, Licenses and Contributors

File:NokiaE50.jpg *Source*: http://en.wikipedia.org/w/index.php?title=File:NokiaE50.jpg *License*: unknown *Contributors*: Original uploader was Notmicro at en.wikipedia

Image:Nokia e50.jpg *Source*: http://en.wikipedia.org/w/index.php?title=File:Nokia_e50.jpg *License*: unknown *Contributors*: User:Asimzb

Image:Gulf E50 Metal Black.jpg *Source*: http://en.wikipedia.org/w/index.php?title=File:Gulf_E50_Metal_Black.jpg *License*: unknown *Contributors*: Original uploader was G patkar at en.wikipedia

File:Nokia wordmark.svg *Source*: http://en.wikipedia.org/w/index.php?title=File:Nokia_wordmark.svg *License*: unknown *Contributors*: Apalsola, Bencmq, ELeschev, Editor182, Hautala, Yarl, 1 anonymous edits

File:Nokia HQ.jpg *Source*: http://en.wikipedia.org/w/index.php?title=File:Nokia_HQ.jpg *License*: unknown *Contributors*: User:-Majestic-

Image:Fredrik Idestam.png *Source*: http://en.wikipedia.org/w/index.php?title=File:Fredrik_Idestam.png *License*: unknown *Contributors*: -Majestic-, Apalsola, Jpk, Martin H.

Image:Leo Mechelin (cropped).png *Source*: http://en.wikipedia.org/w/index.php?title=File:Leo_Mechelin_(cropped).png *License*: unknown *Contributors*: -Majestic-, Martin H.

File:Nokia 150 and nokia 1100.jpg *Source*: http://en.wikipedia.org/w/index.php?title=File:Nokia_150_and_nokia_1100.jpg *License*: unknown *Contributors*: user:Lvova

File:Nokia booklet 3g-10 (3949263497).jpg *Source*: http://en.wikipedia.org/w/index.php?title=File:Nokia_booklet_3g-10_(3949263497).jpg *License*: unknown *Contributors*: http://thenokiablog.com/Reposted by Mark Guim from United States

File:Nokia evolucion tamaño.jpg *Source*: http://en.wikipedia.org/w/index.php?title=File:Nokia_evolucion_tamaño.jpg *License*: unknown *Contributors*: Jorge Barrios

File:All 9xxx.png *Source*: http://en.wikipedia.org/w/index.php?title=File:All_9xxx.png *License*: unknown *Contributors*: User:-Majestic-

File:Nokia E55 01.jpg *Source*: http://en.wikipedia.org/w/index.php?title=File:Nokia_E55_01.jpg *License*: unknown *Contributors*: James Nash

File:Nokia N8 (front view).jpg *Source*: http://en.wikipedia.org/w/index.php?title=File:Nokia_N8_(front_view).jpg *License*: unknown *Contributors*: Editor182, X-Pilot

File:Nokia N900-1.jpg *Source*: http://en.wikipedia.org/w/index.php?title=File:Nokia_N900-1.jpg *License*: unknown *Contributors*: User:Ilya Voyager

File:Nokia E90 communicator.JPG *Source*: http://en.wikipedia.org/w/index.php?title=File:Nokia_E90_communicator.JPG *License*: unknown *Contributors*: User:Georgy90

File:Nokia5800xpress.png *Source*: http://en.wikipedia.org/w/index.php?title=File:Nokia5800xpress.png *License*: unknown *Contributors*: User:TheAdam0s

File:Flag of Canada.svg *Source*: http://en.wikipedia.org/w/index.php?title=File:Flag_of_Canada.svg *License*: unknown *Contributors*: Anomie

File:Flag of Finland.svg *Source*: http://en.wikipedia.org/w/index.php?title=File:Flag_of_Finland.svg *License*: unknown *Contributors*: User:SKopp

File:Flag of the United States.svg *Source*: http://en.wikipedia.org/w/index.php?title=File:Flag_of_the_United_States.svg *License*: unknown *Contributors*: Anomie

File:Flag of Venezuela.svg *Source*: http://en.wikipedia.org/w/index.php?title=File:Flag_of_Venezuela.svg *License*: unknown *Contributors*: Alkari, Bastique, Denelson83, DerFussi, Fry1989, George McFinnigan, Herbythyme, Homo lupus, Huhsunqu, Infrogmation, K21edgo, Klemen Kocjancic, Ludger1961, Neq00, Nightstallion, Reisio, Rupert Pupkin, ThomasPusch, Vzb83, Wikisole, Zscout370, 12 anonymous edits

File:Flag of India.svg *Source*: http://en.wikipedia.org/w/index.php?title=File:Flag_of_India.svg *License*: unknown *Contributors*: Anomie, Mifter

File:Flag of Germany.svg *Source*: http://en.wikipedia.org/w/index.php?title=File:Flag_of_Germany.svg *License*: unknown *Contributors*: Anomie

File:Flag of Sweden.svg *Source*: http://en.wikipedia.org/w/index.php?title=File:Flag_of_Sweden.svg *License*: unknown *Contributors*: Anomie

File:Flag of France.svg *Source*: http://en.wikipedia.org/w/index.php?title=File:Flag_of_France.svg *License*: unknown *Contributors*: Anomie

File:Nokia logo (1865).svg *Source*: http://en.wikipedia.org/w/index.php?title=File:Nokia_logo_(1865).svg *License*: unknown *Contributors*: -Majestic-, Z10x, 2 anonymous edits

File:Finnish Rubber Works (Nokia) logo 1965.svg *Source*: http://en.wikipedia.org/w/index.php?title=File:Finnish_Rubber_Works_(Nokia)_logo_1965.svg *License*: unknown *Contributors*: -Majestic-, Z10x

File:Nokia arrows logo.svg *Source*: http://en.wikipedia.org/w/index.php?title=File:Nokia_arrows_logo.svg *License*: unknown *Contributors*: -Majestic-, Tim1357, Z10x, 1 anonymous edits

File:Nokia Connecting People.svg *Source*: http://en.wikipedia.org/w/index.php?title=File:Nokia_Connecting_People.svg *License*: unknown *Contributors*: Original uploader was -Majestic- at en.wikipedia

File:Nokian logo.svg *Source*: http://en.wikipedia.org/w/index.php?title=File:Nokian_logo.svg *License*: unknown *Contributors*: Z10x at en.wikipedia

File:Nokia Siemens Networks logo.svg *Source*: http://en.wikipedia.org/w/index.php?title=File:Nokia_Siemens_Networks_logo.svg *License*: unknown *Contributors*: -Majestic-, MBisanz, Melesse, WJetChao

File:Nokian pääkonttori Keilaniemessä.jpg *Source*: http://en.wikipedia.org/w/index.php?title=File:Nokian_pääkonttori_Keilaniemessä.jpg *License*: unknown *Contributors*: J-P Kärnä

File:Nokia-Ovi-logo.png *Source*: http://en.wikipedia.org/w/index.php?title=File:Nokia-Ovi-logo.png *License*: unknown *Contributors*: Korg, Neerajprajapati, Ninja5624, SF007

File:Nokia E52 2.jpg *Source*: http://en.wikipedia.org/w/index.php?title=File:Nokia_E52_2.jpg *License*: unknown *Contributors*: User:Asimzb

Image:Motorola Timeport L7089.jpg *Source*: http://en.wikipedia.org/w/index.php?title=File:Motorola_Timeport_L7089.jpg *License*: unknown *Contributors*: User:Wusel007

File:Group of smartphones.jpg *Source*: http://en.wikipedia.org/w/index.php?title=File:Group_of_smartphones.jpg *License*: unknown *Contributors*: gillyberlin

File:IBM SImon in charging station.png *Source*: http://en.wikipedia.org/w/index.php?title=File:IBM_SImon_in_charging_station.png *License*: unknown *Contributors*: User:Bcos47

File:Nokia 9210.jpg *Source*: http://en.wikipedia.org/w/index.php?title=File:Nokia_9210.jpg *License*: unknown *Contributors*: J-P Kärnä

File:Htc Touch Pro2 Georgy.JPG *Source*: http://en.wikipedia.org/w/index.php?title=File:Htc_Touch_Pro2_Georgy.JPG *License*: unknown *Contributors*: User:Georgy90

File:Original iPhone docked.jpg *Source*: http://en.wikipedia.org/w/index.php?title=File:Original_iPhone_docked.jpg *License*: unknown *Contributors*: Andrew from London, UK

File:Galaxy Nexus smartphone.jpg *Source*: http://en.wikipedia.org/w/index.php?title=File:Galaxy_Nexus_smartphone.jpg *License*: unknown *Contributors*: Faramarz, MB-one, SF007, 1 anonymous edits

File:Global Mobile Applications Store Revenue.svg *Source*: http://en.wikipedia.org/w/index.php?title=File:Global_Mobile_Applications_Store_Revenue.svg *License*: unknown *Contributors*: User:MySchizoBuddy

Image:logo visto.svg *Source*: http://en.wikipedia.org/w/index.php?title=File:Logo_visto.svg *License*: unknown *Contributors*: Sfan00 IMG, Smart Fox, Tom Morris

File:SD Cards.svg *Source*: http://en.wikipedia.org/w/index.php?title=File:SD_Cards.svg *License*: unknown *Contributors*: Tkgd2007,

File:MicroSD MemoryCard 002.jpg *Source*: http://en.wikipedia.org/w/index.php?title=File:MicroSD_MemoryCard_002.jpg *License*: unknown *Contributors*: User:Kropsoq

File:TwinMOS 16gb sdhc.jpg *Source*: http://en.wikipedia.org/w/index.php?title=File:TwinMOS_16gb_sdhc.jpg *License*: unknown *Contributors*: User:Z hosen

File:HP PhotoSmart SDIO Kamera.jpg *Source*: http://en.wikipedia.org/w/index.php?title=File:HP_PhotoSmart_SDIO_Kamera.jpg *License*: unknown *Contributors*: User:Afrank99

File:USB-SD-Cards.jpg *Source*: http://en.wikipedia.org/w/index.php?title=File:USB-SD-Cards.jpg *License*: unknown *Contributors*: User:AssetBurned

File:Colors of SD card.jpg *Source*: http://en.wikipedia.org/w/index.php?title=File:Colors_of_SD_card.jpg *License*: unknown *Contributors*: User:SreeBot

File:SDHC HD C16 32GB .jpg *Source*: http://en.wikipedia.org/w/index.php?title=File:SDHC_HD_C16_32GB_.jpg *License*: unknown *Contributors*: User:ACGFAN

File:SDHC Speed Class 2.svg *Source*: http://en.wikipedia.org/w/index.php?title=File:SDHC_Speed_Class_2.svg *License*: unknown *Contributors*: User:Kizar

File:SDHC Speed Class 4.svg *Source*: http://en.wikipedia.org/w/index.php?title=File:SDHC_Speed_Class_4.svg *License*: unknown *Contributors*: User:Kizar

File:SDHC Speed Class 6.svg *Source*: http://en.wikipedia.org/w/index.php?title=File:SDHC_Speed_Class_6.svg *License*: unknown *Contributors*: User:Kizar

File:SDHC Speed Class 10.svg *Source*: http://en.wikipedia.org/w/index.php?title=File:SDHC_Speed_Class_10.svg *License*: unknown *Contributors*: User:Kizar

File:Sdadaptersandcards.jpg *Source*: http://en.wikipedia.org/w/index.php?title=File:Sdadaptersandcards.jpg *License*: unknown *Contributors*: Original uploader was Atanasov at en.wikipedia Later version(s) were uploaded by Brybry26 at en.wikipedia.

File:Canon hf100 with memory card.jpg *Source*: http://en.wikipedia.org/w/index.php?title=File:Canon_hf100_with_memory_card.jpg *License*: unknown *Contributors*: User:Mikus

File:Kingston_19-in-1_memory_card_reader.jpg *Source*: http://en.wikipedia.org/w/index.php?title=File:Kingston_19-in-1_memory_card_reader.jpg *License*: unknown *Contributors*: User:King of Hearts

File:Arduino Ethernet Shield (pre-production sample).jpg *Source*: http://en.wikipedia.org/w/index.php?title=File:Arduino_Ethernet_Shield_(pre-production_sample).jpg *License*: unknown *Contributors*: Matt Biddulph

File:8 bytes vs. 8Gbytes.jpg *Source*: http://en.wikipedia.org/w/index.php?title=File:8_bytes_vs._8Gbytes.jpg *License*: unknown *Contributors*: Daniel Sancho from Málaga, Spain

File:Flash memory cards size.jpg *Source*: http://en.wikipedia.org/w/index.php?title=File:Flash_memory_cards_size.jpg *License*: unknown *Contributors*: Afrank99, Ivob, MMuzammils, Moxfyre, Rhe br, Solomon203, Warden, Zxb, 1 anonymous edits

File:SD-microSD adaptor.jpg *Source*: http://en.wikipedia.org/w/index.php?title=File:SD-microSD_adaptor.jpg *License*: unknown *Contributors*: Yoskov

File:SD card pinning.jpg *Source*: http://en.wikipedia.org/w/index.php?title=File:SD_card_pinning.jpg *License*: unknown *Contributors*: User:Afrank99, User:Jaho

File:Mini SD card pinning.jpg *Source*: http://en.wikipedia.org/w/index.php?title=File:Mini_SD_card_pinning.jpg *License*: unknown *Contributors*: User:JohnHind

File:Micro SD card pinning.jpg *Source*: http://en.wikipedia.org/w/index.php?title=File:Micro_SD_card_pinning.jpg *License*: unknown *Contributors*: User:JohnHind

File:Sd insides.png *Source*: http://en.wikipedia.org/w/index.php?title=File:Sd_insides.png *License*: unknown *Contributors*: User:Saltmiser

File:SD-extreMEmory 2GB alt innen.jpg *Source*: http://en.wikipedia.org/w/index.php?title=File:SD-extreMEmory_2GB_alt_innen.jpg *License*: unknown *Contributors*: User:Manorainjan

File:SDHC inside.jpg *Source*: http://en.wikipedia.org/w/index.php?title=File:SDHC_inside.jpg *License*: unknown *Contributors*: User:George Shuklin

File:Sd4gy crop.jpg *Source*: http://en.wikipedia.org/w/index.php?title=File:Sd4gy_crop.jpg *License*: unknown *Contributors*: User:Wirepath

Image:Activesyncicon.svg *Source*: http://en.wikipedia.org/w/index.php?title=File:Activesyncicon.svg *License*: unknown *Contributors*: Will Taylor (Vector Converter)

File:Activesyncnew.png *Source*: http://en.wikipedia.org/w/index.php?title=File:Activesyncnew.png *License*: unknown *Contributors*: Althepal, Dream out loud, Koman90

CPSIA information can be obtained at www.ICGtesting.com
Printed in the USA
LVOW101533270812

296157LV00005B/129/P

9 786200 715135